SCIENCE, SOUL, AND THE SPIRIT OF NATURE

SCIENCE, SOUL, AND THE SPIRIT OF NATURE

Leading Thinkers on the Restoration of Man and Creation

Irene van Lippe-Biesterfeld
with Jessica van Tijn

Bear & Company
Rochester, Vermont

Bear & Company
One Park Street
Rochester, Vermont 05767
www.InnerTraditions.com

Bear & Company is a division of Inner Traditions International

LIBRARY OF CONGRESS CATALOGING-IN-PUBLICATION DATA

Lippe-Biesterfeld, Irene van.
Science, soul, and the spirit of nature : leading thinkers on the restoration of man and creation / Irene van Lippe-Biesterfeld with Jessica van Tijn.
p. cm.
Includes bibliographical references
ISBN 1-59143-055-0 (pbk.)
1. Human ecology—Philosophy. 2. Philosophy of nature. 3. Gaia hypothesis. I. Tijn, Jessica van. II. Title.
GF21.L56 2005
304.2—dc22

2005017021

Printed and bound in the United States by Lake Book Manufacturing, Inc.

10 9 8 7 6 5 4 3 2 1

Text design and layout by Virginia Scott Bowman
This book was typeset in Sabon and Avenir with Basilea and Optima as the display typefaces

Human beings are a part of a whole called by us the "Universe," a part limited in time and space. We experience ourselves, our thoughts and feelings, as something separated from the rest—a kind of optical delusion of consciousness. The delusion is a kind of prison for us, restricting us to our personal desires and to affection for a few persons nearest to us. Our task must be to free ourselves from this prison by widening our circles of compassion to embrace all living creatures and the whole of nature in its beauty.

ALBERT EINSTEIN

CONTENTS

FOREWORD

When three thousand of the world's leading experts on protected areas and nature conservation met in Durban, South Africa, in September 2003 at the World Parks Congress, they may well have expected to address technical issues of nature management, especially how to minimize the impact of humans on ecosystems. Imagine their surprise to find numerous ceremonies throughout the event that underlined the unity, rather than the divide, between people and nature.

Particularly remarkable was the active participation of many groups of indigenous peoples from all parts of the world. Most of the world's protected areas have been established on lands that are considered ancestral homelands by these indigenous and tribal peoples, who have long had a spiritual relationship with parts of the landscape. Indeed, such sacred sites are still found in all parts of the world, from Sweden to Swaziland. The presence of this wonderful cultural diversity in Durban forced the people attending the Parks Congress to reexamine their own motivations for working in protected areas. Yet, as Princess Irene stated so clearly in her address at the congress, it is necessary that not only indigenous peoples live the deep connection to all life; it is necessary that all humankind rediscover its connection to the universe and the Earth herself, for this will help us to redefine our humaneness and see our place as part of the life community. That we are part of the symphony of life is what we must know for the future. Are the parcels of land we protect set aside merely to save them from worst excesses of modern consumer society, or are these national parks and protected areas instead established as an expression of our modern culture—that is, out of our respect for what these outstanding natural places can contribute to our national identities? The answer to this question is no trivial matter, nor is a single answer possible. For many of the three billion people now living in cities, nature is a destination, a place to be visited on vacation, accessed by road, captured on film, and experienced in small bites—there we may still find a hot shower, a four-course meal, and a warm bed at the end of the day. Yet for the other three billion people who actually live among nature, this separation is not nearly so clear, and the unity between people and the rest of life may be more apparent.

Many people today are "Earth lovers," as we learn from the perspectives presented in this book. While scientists, including the ten thousand or so experts within the six World Conservation Union (IUCN) commissions, help us to understand the living world, this understanding is not enough to change the way we behave as a modern society. Instead, we need to join science with emotion and recognize that our conservation efforts will be successful to the extent that we can enhance, support, or build anew the emotional relationship between people and the rest of nature. Conservation, after all, is built on caring for the Earth.

Interestingly, modern science is also generating new evidence that human altruism is a powerful force and may be unique in the animal world, providing scientific evidence that our emotional ties to nature have adaptive value. Research at the University of Zurich, published in 2003, showed that a minority of altruists—people who are willing to work for some positive goal without demonstrated self-interest and without any likelihood of material gain—can convince a majority of selfish individuals to cooperate. The interaction between altruists and egoists, or, to use the terms of this book, between lovers and haters, has been an important impetus for human cultural evolution. And as this book demonstrates, if we are able to mobilize sufficient numbers of Earth lovers, then we can enter the twenty-first century with a new sense of confidence in the ideals of sustainable development.

Yolanda Kakabadse
Former president of the World Conservation Union
Gland, Switzerland

INTRODUCTION

Twelve people, twelve visions. I contacted twelve individuals from all continents who, either professionally or based on their personal experiences, derive inspiration from the same source: their deep affection for the Earth and society. These twelve represent a wide variety of disciplines. Despite their differences, however, the theme of this book—the relationship between humankind and nature—is an essential issue that matters to all of them in their everyday lives. All of those interviewed are original thinkers who have followed their own insights and formed their own opinions about the relationship between humankind and nature.

Years ago, I thought I was alone in my questions and discoveries about our place in the natural world, and launched a quest for kindred spirits who were active in this arena. In my search, I learned that people around the globe have given the subject serious consideration at different levels and in different ways. This encouraged me to speak with a number of them and to share my findings in this book.

Reading their work and reviews of their work over the past decade has confirmed my own perceptions and has helped me continue my own quest. Each of them represents an individual color in the broad spectrum of the theme of this book, like stones in a kaleidoscope that form new patterns with every turn.

Those interviewed here reveal a deep respect for all efforts in the field of nature preservation, but despite our progress, they believe we are ready for the next step. Perhaps we should call it an additional awareness that we are connected to nature both biologically and spiritually: that we *are* nature. Terms such as *supervisor* and *owner* pertain to an anthropocentric paradigm of our relationship to the natural world—a paradigm leading to a way of thinking and acting that, alas, is not extinct. The consequences of this view have been horrible, as we will read in Gareth Patterson's story in the pages that follow. Within this paradigm, humankind is in charge and nature is mere matter—chattel— with which we can do as we please. What is more, this way of thinking suggests that we are separate and distinct from the natural world, rather than part of her being.

Our attitude toward nature has changed considerably over the years. Many of us today regard ourselves as its trustees or caretakers. Though this might have a more appealing ring than *supervisor* or *owner,* it still does not allow for our unity with the Earth. Yet increasing numbers of people do feel they are a part of nature—and surprisingly, many of them have experienced what I call a "magic moment": a moment of unity with all life. I have had such an amazing experience myself, and it changed my attitude toward the Earth. I felt the sense of being one with all life around me—the separation vanished. Perched high up on a Swiss mountain, I saw through everything and felt and heard through it all as well. I was part of the waving blades of grass, the colorful flowers, the trees, the rocks, and the mountains. I was aware of having converged with everything alive and of being united with nature's exuberant, joyful celebration, its powerful yearning toward life.

Since then, my love for the Earth has led me to become a student of life.

Based on my experiences, I have drafted some basic questions that I regard as the core of this book. These questions have preoccupied me, and I was curious how those I interviewed would respond: What is nature? What is our human relationship with it? Why have we become disassociated from nature? How does nature affect the life of each of us? How do we experience the unity of life in everyday reality? Is there a balance between what we call *science* and our personal experience? Can and should science provide answers to these questions? Can we manage without science? Can an empirical approach to nature and one based on human, sensory experience complement each other? How do *reason* and *feeling* relate? How do people from different cultures and disciplines answer these questions? As you will see during these conversations, the central questions become less important than the line of thought that emerges from them. No easy answers are available, but I was not looking for them. My main concern has been to begin a dialogue that can encourage deeper thought and investigation on the part of each person reading this book.

My central question to those interviewed here is this: What is love?—for these twelve have manifested a relationship with the Earth that is based on love, and love for all life here is what each of us must grow to feel if the Earth and her species are to survive. We often restrict our use of the term *love* to the idea of exclusivity or ownership, which arises from our fear of losing. In some cases, a sense of loneliness and insecurity causes us to cling to people and objects, making them possessions. If we reconnect with nature, and with the nature inside us, this feeling may disappear, for we are not alone. In fact, we are connected to all life on Earth.

If we learn to be receptive to the life-forms around us, we will discover subtle energies: the light, the essence present in all that lives, the love that flows through everything that exists. This connection, formed with our heart, helps us realize that we are surrounded by love and allows us to feel supported. The answers I received to the question "What is love?" were both surprising and heartfelt.

In his book *De natuur als beeld in religie, filosofie en kunst* (Nature as Figure in Religion, Philosophy, and Art), Professor Matthijs Schouten notes that nature can be viewed from various perspectives: from the point of view of a person who perceives without feeling—that is, someone to whom a primrose is but a primrose, for which he feels no love; or as a person who perceives through feeling—that is, someone to whom a primrose is a star, a sun, a fantastic image, anything but "just" a primrose; or as a Zen master, who perceives the little plant in a way that enables the plant to reveal itself, who feels that this humble little plant unveils the entire mystery, the mystery that is the source of everything and that life embodies.

Since the beginning of human life on Earth, we have struggled with our relationship with nature. We have loved nature, hated it, used it, romanticized it, and seen it as a purely mechanical entity. Clearly, we are starting to understand that our anthropocentric approach has ultimately led us to deplete the Earth. What, then, can we do?

Actually, we need to change our manner of thinking and our awareness, rather than simply our actions. In the pages that follow, Patricia Mische recommends going one step further and transforming our political thinking with a more acute awareness of the greater global society.

I aim to show here that everything relates to everything else: our solidarity with or seclusion from all that lives, poverty, health problems, climate changes, and natural disasters. And all of it begins with our awareness—or our lack of it. Key is the point that nature will benefit if people feel a sense of solidarity with it. Becoming aware of the interaction among all life-forms can be a deeply spiritual experience. The web of life in such fragile balance, its beauty and synchronicity, is overwhelming the moment we are receptive to it. Patricia Mische refers to this receptivity as *Earth literacy,* "the ability to read and write the Earth." With this ability, the world becomes a miraculous place in which to live. Rupert Sheldrake keeps rediscovering this miracle, and Hans Andeweg teaches us to listen to it. Professor Arne Naess discusses *ecosophy,* ecology combined with deep-earth philosophy. Gareth Patterson shares how he learned to understand the unity of all life thanks to lions. Denise Linn believes that every individual can reach deep within himself or herself to find long-sought answers to life questions. Patricia Mische adds that "if we lack an alliance with nature, we lack an alliance with ourselves." Jane Goodall explains how in the wilderness, she feels closer to spiritual power.

Thus, we can place ourselves in the web of life as we see fit, provided we understand the power of our thoughts, for they can destroy or repair the quality of life. Masaru Emoto shows this with his astounding images of water crystals and tells us, "You need to approach water with your heart." His words remind us that our respect directly affects both ourselves and the life-forms around us. Understanding this, Rigoberta Menchú Tum has dedicated her life to peace, offering the message that there is an alternative way of life, which she bases on her own ancient Mayan culture.

The quality of our life is linked with all life on Earth and beyond. This gives each of us a unique, personal responsibility. The knowledge that we are co-creators of our life on Earth, that our ideas and actions have consequences well beyond their scope, could serve to scare us and make us withdraw; but it might also enhance our awareness and add meaning to our life. The knowledge that we are creating our own reality makes life a wonderful learning process, in spite of and with all its difficulties. This in turn may give rise to new priorities with respect to society as a whole.

We sorely need these new priorities. Credo Vusamazulu Mutwa is deeply concerned about the loss of dignity among black people, seen in the complex and harsh reality that surrounds him in his part of Africa. James Wolfensohn, whom I interviewed during his tenure as president of the World Bank, sees a need for universal equality, has aimed to contribute to this cause through his work, and sees those who incite hatred as reprehensible.

All those I spoke with emphasized that neglect of our surroundings will lead to neglect of ourselves, and that our acts of destroying will cause our self-destruction. It is a matter of choice. We can choose our actions and our course. Choices become easier once we understand our position in the web of life. A new mind-set is ascending.

I spent a year on the road with Jessica van Tijn, a journalist and personal friend, in search of these twelve remarkable people. Most are pioneers in their own disciplines, aware of the consequences of publicizing their personal discoveries and driven by an inner curiosity about life. All of our conversations were vivid, fascinating, and inspiring. These written accounts can convey only a fraction of the full substance of great themes we addressed, let alone of the people themselves. And yet my hope is that they will inspire your self-discovery. The bibliography at the end of each interview may further this effort. *Science, Soul, and the Spirit of Nature* is an invitation to become acquainted with this group of professionals through their special stories and insights. They have all dedicated their lives to searching, experiencing, studying, and sharing their findings. Interestingly, although all are well known, most did not know the others, yet revealed here is a fascinating, surprising, and somehow necessary relationship among them. Their individual differences only reinforce their common wealth of ideas and feelings, their love for our Earth.

May this book help inspire you to love the Earth and to find your way as co-creators of our common future, which is in your hands.

RIGOBERTA MENCHÚ TUM

None of us would start anything—
a congress, a new job, a new business,
an event—without greeting the sun
and using its energies.

The weatherman has announced that an exceptional solar eclipse will take place at 5:30 in the morning, a onetime occurrence in this century. My appointments with Rigoberta Menchú Tum, Mayan winner of the 1992 Nobel Peace Prize at the age of thirty-three, have been canceled three times because of her frequent travels. But today we will finally meet in Berlin. For more than a decade, she has traveled all over the world with seemingly inexhaustible energy on her mission to speak in support of peace.

I am particularly interested in hearing about her connection with the sun, for the ancient Mayan culture, through its calendar, is intimately intertwined with this celestial body. On my way to the airport, I suddenly see it: the enormous red-orange sickle of the rising sun, with its greatest portion still concealed by the moon. The dazzling radiance through the misty morning sky is breathtaking. I feel irresistible joy as I gaze at this miracle, without being able to or wanting to understand why.

Rigoberta is in Berlin for an ecumenical conference. Our flight arrives so early that we have time for a stroll through a public park behind the hotel. The branches of old trees hang over the water of a large pond. Jessica points to the sign near a gnarled old plane tree—a *Platanus acerifolia*—right beside me. The text, intended to interest children in nature, impresses me and reminds me of a quote posted in the park where we walked before our interview: *"Hör doch mal! Auch Bäume können sprechen. Die 'Sprache' der Bäume kann man nur erfahren, wenn man ihnen genau zuhört. Lege Dein Ohr ganz dicht an den Stamm. Was kann man hören? Kommen alle Geräusche, die Du hörst, vom Baum?"* (Listen! Trees can speak, too. To hear the "language" of trees, you must listen very carefully. Place your ear close to the trunk. What do you hear? Are all the whispers you hear coming from the tree?)

We find Rigoberta in her hotel and are met by Alfonso Alem Rojo, executive director of the Foundation Rigoberta Menchú Tum (FRMT), in Mexico. He joins us during our conversation. The sun-splashed terrace is a wonderful place to sit and talk amid lush greenery. We speak Spanish because Rigoberta does not speak English. "All I can tell you in English are a few small jokes," she explains with a twinkle in her eyes. We get started, happy to see each other again after our previous meetings, when she had shown me dozens of photographs of mass graves in her native Guatemala on one of her visits to raise awareness of the atrocities that have occurred there.

This tiny, courageous woman from the village of Chimel in the province of El Quiché encountered death and the evil of humankind very early on. First, her brother, Patrocinio, was murdered. Soon after, her father was killed—burned alive—when the army stormed the Spanish embassy in Guatemala. He and a group of Indians from El Quiché had asked the Spanish ambassador for help in ending the repression in his country. Rigoberta's mother was then kidnapped, severely tortured, and killed as well, even though the official data suggest that she is still alive. Because none of the remains of these family members have ever been found, Rigoberta has been unable to give them a decent burial.

Soon after losing her family, Rigoberta fled Guatemala for Europe, where she tried to call attention to the plight of her country. Her overt activism rules out returning to Guatemala. In 1983, she published her autobiography, *Me llamo Rigoberta Menchú,* with the help of the anthropologist Elizabeth Burgos. In 1992, she was awarded the Nobel Peace Prize. Her whole life has changed since then, as she tells us. One aspect remains the same, however: Just as before receiving the award, she travels all over the world to speak about peace and justice—and to tell people about traditional Mayan culture and the importance of the wisdom of the ancient tribes for our world, for the Earth.

Rigoberta lives in Mexico City now because her own country is too dangerous for her to do her work. This is but one of the many sacrifices she makes to fulfill her mission in this life. Her fragmented experience of her son Mash's childhood is another. She is fortunate that she can rely on her husband, Ángel, a Guatemalan like herself who also lost members of his family during the country's dark years.

I am delighted that despite all the canceled appointments in the past year, we meet today so that Rigoberta can join the circle of people in this book.

What is nature?

Rigoberta Menchú Tum: To me, nature is the convergence of lives, of eras, and especially of people, who together form a reference to life. We would not be "life" if we did not share with the other lives on Earth. The Maya are in close contact with cosmic energies, and in this respect the energies of the mountains, the forests, and the rivers are undeniably part of our own energy. This [cosmic] energy cannot flow without *their* energy. I am referring not only to the energy of the living but also to the energy of the dead.

We Maya schedule our lives according to our own calendar, which consists of twenty Nlawales, which each repeat thirteen times in three hundred sixty days. Each day has a Nlawal, that we thank and honor, and which helps us with its positive energy and makes the sun rise and set and the wind pick up and subside. Nobody seems to wonder why the wind picks up or subsides. But the Maya take an integral view of life that includes these patterns as part of nature. The struggle for the environment is difficult for us to understand, considering that the Spanish word for *environment (medio ambiente)* literally means "half surroundings." To us, something as important as the environment cannot possibly be "half"! The environment is integral, and each of us can evolve with greater strength and greater reception of the energies, or with less. In either case, elements of the Nlawal help us live in the cosmos.

During our childhood, for example, our parents teach us to write down our dreams and to remember them and talk about them. So if we have a dream, we tell our parents, because we do not always understand all our dreams. Over time, we obviously learn to understand far more about our own dreams and their meaning. At the same time, we learn to grasp the greater dimension of nature and of all living beings around us. Eventually we become highly sensitive to changes in the world and the cosmos. This makes us far more receptive than other cultures to these things. Our greatest support by far, however, is our own calendar, which goes back thousands of years.

Given that you spend so much time traveling to all corners of the world, how do you cope with all those time zones, time changes, and appointments on dates according to the Gregorian calendar? Most places you visit do not observe the Mayan calendar.

RMT: A day is a day, and sometimes you even forget the exact time and date. Of course, those two calendars [Gregorian and Mayan] do not correspond. At this moment, for example, I could not even tell you the exact date according to the Mayan calendar. My agenda is ruled by the Gregorian calendar. On days or moments that I dream something remarkable or have a "perception," or if I feel lethargic or become interested, I can always check our calendar to search for the exact date and its significance. Our calendar usually offers an explanation for my feelings or experiences.

I feel that I always keep the "time"—the time according to the Mayan calendar—even if I am not always aware of the date. And you need not be a specialist to understand this calendar.

So the Mayan calendar is still important to the Maya?

RMT: Absolutely. There are three stages that connect each of us to this calendar: one, the moment we are conceived by our parents; two the moment of birth; and three, the "rhythm of the time" between one and two, which determines our behavior.

Each Nlawal has both positive and negative attributes. We all have both within us, and whether we cultivate the positive or the negative ones within ourselves is up to us.

How would you translate Nlawal?

RMT: The spirit, the universe. What makes each individual unique. You could call it a cosmic DNA.

Is it every individual's relationship with the cosmos, sort of* la cuerda unica*—your unique cord that connects you with the cosmos?

RMT: Yes, and that determines the course of our lives. It is beautiful, although some people have a superficial interpretation of the indigenous cultures. They condemn our way of thinking, of living, and of believing. All this is part of our spirituality. For example, I always light a candle when I feel my energies fading, if I feel negative energy that makes me need the elements fire and light. That recharges me with new energy and helps me discard the negative energy.

But other forces are beyond my control. In such cases, I use spiritual guides, who on my behalf ask the cosmos, nature, the mountains, or the heart of the heavens and the heart of the Earth. They mediate for me and on my behalf.

What are those guides called?

RMT: In the West, especially in books, they are always referred to as *shamans*. I believe the Spanish made up that name for everything they associate with natural healers. The Maya have other names for their spiritual guides: In Guatemala, for example, we call them *ajq'ij*.

I remember that the Kogi Indians in Colombia train the "chosen ones," or the* mamas*, who must live in the dark for the first eight to sixteen years of their training. What is the Mayan tradition?

RMT: In our tradition, these guides are born on a special day, at an exact time—on a day known, according to our calendar, as *trece,* which is the strongest symbol of all. Next, these guides need to cultivate their gift and seek spiritual enrichment.

You once said:* Danos útil existencia, *"Make our existence useful." Could you elaborate on that?

RMT: *Danos útil existencia* was the prayer of the first Maya when they were formed and created from yellow and white maize. No tranquil and easy life, but a useful existence, so all positive forces and relations are used.

When our people pray at Mayan ceremonies, something very beautiful happens. None of the elements is overlooked at such a ceremony. We greet the rising sun to ask for light, energy, warmth, and strength. We greet the setting sun to invoke history, to ask our forefathers to give us strength and to stand by us, and to greet small night animals. We greet the wind as it picks up to clear our paths . . . When the wind dies down, we call on the energies of Mother Earth and of the seeds, so that we may blossom again, as seeds do. And when we appeal to the heavens, it is to protect ourselves in the universe, in the Void. The heart of the Earth is like our mother, a mother who nurses her children, who nurses life.

During the ceremony, we take all these elements to form what we call the Mayan Cross, which uses the six elements for the six energy points. Aside from the North, South, East, and West, symbolized by the sun rising and setting and the wind picking up and subsiding, at the top is the symbol of the cosmos, or the Void, and at the bottom is the symbol of the heart, or the Earth. This makes for a three-dimensional cross.

I associate the present lack of equilibrium in the world with the way we deal with the Earth and the fact that we have strayed so far from her . . .

RMT: I am under the impression that relationships between individuals are deteriorating, primarily because of a lack of mutual respect. It seems as if the unspoken rules of interaction are disappearing, which is causing us to lose track of everything. I am convinced that a proper social equilibrium is possible only if we respect each other. If the main purpose of our laws is to punish violations and offences and little is done to prevent them from happening and to avert a loss of respect for others, then the writing is on the wall.

The Mayan tradition emphasizes peace. *Peace* is not synonymous with *absence of war*. Peace is referred to as *utzil*, which basically means "living with goodness." Living with goodness is based on a code of conduct, an ethical code that is based on respect. Loss of respect shifts the entire balance.

The problem is not so much that lack of respect quickly leads to offending people; that is the least of the problem and can be solved through a change of attitude. At the moment, however, I notice lack of respect in so many different situations, whether we are referring to the attitude of young people toward their elders, the way we interact with nature, or our mutual relations. Much has already been lost. I believe that the loss of standards and values is a universal trend that affects all of humankind. We need to find a new way to implement those standards and values again, for a society cannot focus exclusively on punishment rather than prevention of as many problems as possible.

Human history is filled with wars and changes in the balance of power. Even during the heyday of the Maya, life was not entirely peaceful and wars were waged. At any rate, Mayan culture has always assumed our spirituality to be the underlying principle rather than an ultimate objective.

As evidence of Rigoberta's deep belief in the importance of respect, the Foundation Rigoberta Menchú Tum has drafted a so-called code of ethics to emphasize how much Rigoberta and the people who work with her value peace and equality in the world:

> *Peace cannot exist without justice.*
> *Justice cannot exist without fairness.*
> *Fairness cannot exist without development.*
> *Development cannot exist without democracy.*
> *Democracy cannot exist without respect for the identity and dignity of cultures and peoples.**

*From *Rigoberta Menchú: Una década de compromiso por la paz y la justicia.*

Following the attacks in the United States on September 11, 2001, you said you were surprised that the entire world spoke about nothing else, that so many more people were killed during the dictatorship in Guatemala but nobody talks about them.

RMT: Yes, I am convinced that the dead in the First World are mourned far more vociferously than the dead in the Third World or in poorer countries. I remember very well that seven thousand people died during the invasion in Panama. The world took little notice, and the stock exchange was certainly not affected! Two hundred thousand have been executed in Guatemala, and about fifty thousand are still missing: family members, brothers and sisters, acquaintances. Officially, however, they are still alive. All those figures are not enough to shame or frighten the world. The worst part may be that there are now about three thousand mass graves that we have more or less traced. To think of all those unidentified human remains lying there without achieving an impact on the world . . .

When I commented on the responses to the attacks on September 11, I also considered the fourteen million people who starve to death every year in this world—and all those millions of people with cancer or AIDS who have no access to proper medical care, and the number of children who die, day after day, from curable diseases and infections. It is inexcusable that a few terrorist attacks—do not get me wrong: I am firmly opposed to such forms of terrorism—are diverting public attention from all these other problems. We need to change the global perspective that overlooks the suffering and deaths of so many millions of people. After all, that is an invisible form of terrorism.

According to NATO figures, one hundred thirty million children live on the streets, largely orphans or kids abandoned or abused by their parents, working since childhood without any prospect of education or improvement. How can we explain to them that this is the world in which they live? All these signs of imbalance in the world must affect humankind, must virtually destroy us.

Years ago I believed in world justice. I thought that in the past half century humankind had achieved consensus and a dialogue based on exchange. Now, I no longer believe in that and see the survival of the fittest everywhere. Nobody seems to hear anymore all the millions whom I am talking about.

In my work I try to surround myself with "regular" people. I also meet many heads of state and government leaders. I still search for leaders who maintain their standards and values in spite of everything. But I rarely meet them. That is why each of us has to accept his or her own responsibility, so that we do not entrust our fates to a few insane people who think they are the Messiah. I know a few of those heads of state who truly believe that they are the Messiah. They know nothing at all about life. Their arrogance must have gone to their heads.

What lies ahead for humankind? I do not know. All I know is that every day brings more pollution, more destruction. How will we die? From cancer, AIDS, a bomb? From starvation, violence, or loneliness?

Can you die of loneliness?

RMT: Absolutely! Then you continue as a living corpse. You may not actually die.

So many people feel they are alone in this world. If only they could find solidarity and comfort, and if only they could be heard.

RMT: We cannot accept the sadness and loneliness. We need to keep believing in a world that is more honest, more human, and more balanced. We must try to achieve a more peaceful coexistence and to lead a good life. We need to realize this Utopia. If we let ourselves be tempted to adopt a negative attitude in life, that would be a sign of overall failure in the world. That is how I feel.

Does your culture help you strike a spiritual balance?

RMT: Absolutely, even though many of us often forget how important spirituality is. We forget to stay in touch with our forefathers. In my case, living with all those concepts and principles is all very well, but when your parents and your brother are murdered . . .

One of the ways the genocide in Guatemala affected me was that I no longer believed anything or anyone. Should I believe that there is a God, and that there is a global equilibrium? Over time, my husband, Ángel Canil, and my son, Mash Nawaljá, have helped me regain my faith in goodness and in the importance of spirituality in life.

Rigoberta is wearing a beautiful necklace shaped like a sun, depicting a Mayan calendar on the back. I noticed it when we met earlier this morning.

What does the necklace symbolize?

RMT: I live with the sun and the Mayan calendar.

I tell Rigoberta about the solar eclipse early this morning. And I mention how remarkable it seems that we finally meet her today, on the day of the solar eclipse, knowing how important the sun is in her life and culture.

It was so moving that I cried tears of joy. What did the solar eclipse mean to you?

RMT: That the sun and the moon love each other.

Now I understand why I was filled with joy.

RMT: (Smiling) The sun is important in the Mayan Cross, with its six cosmic points, that I described. The sun is connected to two of those reference points. Our forefathers calculated the time with tremendous accuracy based on their cosmic knowledge. For that reason alone, our ties with the sun and the other cosmic elements are incredibly strong.

None of us would start anything—a congress, a new job, a new business, an event—without greeting the sun and using its energies. These energies are vast and inexhaustible.

Our relationship with the sun is so close that when we grow, we believe that our bones feed on solar energy. I believe that we are like plants: Plants raised without sun grow very differently from those that soak up energy from the sun every day. This happens with people too, for we are also part of nature.

Do you live with an awareness of the sun every day?

RMT: Yes, I even apply cream every day to protect my skin from sunburns!

I believe that the presence of the different elements and nature in people's imaginations is a good thing. Thinking about the depth of the sea never hurt me. Taking the power of the sun and the other elements into account is very important.

All morning the sun has been shining brightly, but the elements seem to have been listening to our conversation, and rain starts to fall on the hotel terrace. As the drops fall more heavily, they force us to move to a table inside. Rigoberta continues her story.

RMT: I want to gather all tales about the sun. In our tradition we refer to Grandpa Sun and Grandma Moon, who always figure in the stories and fairy tales our grandparents tell us. This reinforces our close relationship with these elements. The solar eclipse early this morning means that today is a sign of new lives and new times. Once I am back in Mexico, I will check today's date according to the Mayan calendar to determine its significance.

Some time after our visit, we receive information on its date from Rigoberta's office in Mexico City. The day we met was the ninth of Q'anil, according to the Mayan calendar, which is a day with a high energy level. Q'anil symbolizes vitality and fertility. This day is associated with plants and living creatures that reproduce, while nine represents completion of something in life and expansion to a new stage. My conversation with Rigoberta was the last to be completed for this book. In keeping with the ninth of Q'anil, it is my hope that this volume will be fertile ground for new ideas and actions.

Rigoberta continues her explanation about the function and importance of the Mayan calendar.

RMT: The calendar tells us when to ask for things and when to give thanks. It also indicates which days are right for performing business transactions or making new friends or entering relationships. All of us have wished, at one time or another, that we had done something yesterday or that we had waited another day, because today is not at all as we had intended!

The effect depends on how receptive you are to the energies. It is a matter of intent and faith, life's spirituality. When somebody invokes the energies of light with such force, you will feel that too. We often use fire, especially in our ceremonies. Fire is a wonderful indicator of ambience.

Is the effect also related to the attitude of a group of people present? I mean, if the entire group is receptive, does that make the effect different or more pronounced?

RMT: Alfonso Alem answers the question. He explains what happened when the different directors of the Foundation Rigoberta Menchú Tum met in January: "Like at the other gatherings, several spiritual guides attended. Most of our meetings last a week. Throughout that week, we arranged an altar in the middle of the table where we held our meetings, in keeping with the ancient Mayan custom. I do not come from a Mayan background and was deeply impressed with the use of the fire. We all noticed how the fire almost went out when we discussed complicated issues. We had to work even harder, with ever-rising dedication and conviction, to keep the fire going. You could see the glass around the candles blacken at those moments. Whenever we resolved an issue or switched the topic of conversation, the fire would flare up. So I am absolutely convinced that the elements are very sensitive to the mood of their surroundings. . . . That is why it is such a pity that some people fail to distinguish material matters from spiritual needs."

What is our place in this world? What is the role of humankind?

RMT: I believe that the more challenges people encounter in life, the more they revert to spirituality. Sometimes this happens very slowly. We do not suddenly stop and realize what we are up to and decide at that same moment to open ourselves up. It takes time.

This time, however, is a perfect opportunity to reflect on life, on its scope and the importance of every one of us. It gives us a chance to cultivate self-confidence in our children. I believe we need to focus on our children at this time. The elderly are far more difficult to influence.

Three things matter: We have the past, which is already made. Then there is the present, which we make, for better or for worse. What really matters, though, is the future, which is still intact in everybody who has yet to be born. We need to hand down and return as much as possible to future generations. This is the main reason why I am writing fairy tales for children. Sometimes all my painful memories make going back in time difficult. But I hope that this way we can at least protect the next generation and offer them a new equilibrium.

So many other people are working with children now. Many of the people we have met for this book see this as an important and necessary course of action. It seems as if children today are far more receptive and are more open-minded when they enter the world—as if they want to show us the way . . .

RMT: We Maya believe that 2012 will be a year of major, decisive change. According to our calendar, such a change occurs once every fifty-two hundred years, when the old era ends and a new one begins. I expect the gateway to this *baktún** to be bloodstained. The present year corresponds with *tijax* on the Mayan calendar, which symbolizes the onyx knife. This is also the symbol of the natural healers and physicians, as well as the symbol of violence. Since the beginning of the year, we have focused more on things that may cut life short: car accidents and other calamities, disease, natural disasters, wars. The other symbol for this year is *ajaw,* regarded as the formative power and architect behind this era. Considering the events of the recent past, it corresponds roughly with the following symbols: the outbreak of SARS, the second Gulf War, and earthquakes. They are all related to the approach of

*A *baktún* is a period of forty years. The period of 5,200 years (130 x 40) is critically significant in both the Mayan calendar and the counting system. This period ends in 2012.

2012, which means that people have to make a special effort to prepare for this major change and have to struggle to survive. It also forces us, before we enter this new stage, to reflect on what we have done with the world and to take a look at our inner selves.

It is interesting that some kind of divide is appearing. On the one hand, people feel increasingly drawn to the spiritual aspect of life. On the other hand, there are more and more wars and ominous trends.

RMT: In Guatemala, young people are preparing. Many want to study law or medicine, or they want to become engineers. If you think about it, these are the professions where demand will be greatest. We need people to dispense justice, to keep us or make us healthy, and to help build a new society.

And what about plans to open the first Mayan university?

RMT: We still plan to establish a university in Guatemala that will offer such knowledge, experience, and tradition. We have organized three congresses, and the Academia de Lenguas Mayas already exists in Guatemala. We are now assessing the quality of Mayan education at established schools: Where is the education bilingual, which systems are used? We want to reach all twenty-four thousand teachers already involved in Mayan education and include them in the comprehensive Mayan school system.

I realize that Rigoberta has not yet answered my question about our role here on Earth. So I ask her again:

What is our place in the community of life?

RMT: We believe that people should act as trustees of the universe. Their responsibility is to provide care and to pave the way toward life and positive energies. The elderly are responsible for providing guidance, because they already have life experience. Wisdom is something that takes a lifetime to develop. We can benefit from the wisdom they have already acquired. Individual wisdom may be transformed into universal wisdom in this way. It is the same with human rights: Collective rights are indistinguishable from individual rights and vice versa.

Do you regard people as the most important life-form in the world?

RMT: Ask the other life-forms!

This is Rigoberta's first response, followed by a perplexed look. She has not really understood my question. But once Alfonso Alem steps in and tries to explain, I realize why: She is unfamiliar with our Western, fragmented approach to considering life on Earth. She then replies.

RMT: How strange to divide the world into physicians who look after our health and to entrust education to teachers, our economy to economists, and culture to cultural experts. Such fragmentation is bad for life, as it inserts an impossible and inconceivable division in all things that should not be regarded as separate. A dentist, for example, cannot look only at teeth and molars and ignore signs emitted by kidneys that reveal a great deal about the problem with his patient's teeth. Likewise, an economist cannot possibly be oblivious to the culture of a country in which he is studying or running the economy.

Many of these differences in outlook arise from the *fortuna* concept. This is interpreted differently in the West than in our culture. Westerners are quick to associate the concept with financial means, savings, and disposable income. To us, *fortuna* also signifies spiritual wealth, the great forces that humankind possesses. This wealth is protected by our spiritual guides, who warn us whenever these riches are threatened. They are not really referring to our economic situation, although that is obviously important as well. Aside from needing to keep body and soul together, I also need to harmonize my energy level with my spiritual health.

Rigoberta, if we agree that the world balance has been upset, how can we restore the equilibrium?

RMT: We need to take concrete steps. Ideally, we should use nature and natural resources, while returning to nature what it needs. This means that we need to invest in nature and restyle our behavior. But there are so many of us here that in the meantime—until we figure out how to behave—millions of trees die every day, and economic prosperity takes precedence over everything else.

I believe very strongly in our work at the foundation and in keeping projects as small as possible. We are struggling to preserve the holy sites of the indigenous peoples as a plan for how to deal with nature. Vast areas are not our primary concern. Sharing a common space matters far more.

This day and age require that we take specific action, and that we share the practices of our Mayan culture with all humankind. Once we open the Mayan university, we hope to enroll non-Maya as well as Maya.

How do you find the energy and inner strength to keep going?

RMT: We live only once! (Joking) But I will live twice. I will return as a coyote in my second life, and I hope there will still be mountains then! . . .

Do you consider yourself a pioneer?

RMT: Not really. In my view we accomplish so little. It takes such a long time to get anywhere.

Do you feel alone in your work?

RMT: Often I do. My work should be so much easier. Borders, for example, should not be an obstacle. Mutual, human contact should be strong enough to allow us to work together toward a common objective. Instead, everything seems to happen with such exasperating slowness. But if there is a will, there is a way. If two parties want to work together, they can achieve miracles.

Do you encounter many obstacles?

RMT: People seem to think that I am a celebrity, and that my every request is granted. That is not really true. I still have to fight to protect the Mayan people and to inform the world about our rights and language.

(Smiling bitterly) If my struggle were not for a collective cause, I would probably be deeply suicidal or would have died of loneliness.

What drives you? Did you begin your pursuit of a better, more honest, and fair world as a child?

RMT: I do not feel the need to search for an explanation. Things are the way they are. But I know that when I return from my travels and see girls who are as I once was . . . I have no idea why the process has become so violent since then. Sometimes I return to my native country and think: Some spirit stole my body and took it away.

Do you feel that way because of all the injustice?

RMT: Of course. Sometimes you wake up one day and suddenly long to understand and to increase your knowledge. The human mind is like a sponge that keeps absorbing new things. I absorbed knowledge, ideas, and the concept of time, which helped me understand the world. All this happened in an important, decisive period in Central America, with great thinkers, writers, critical voices, people who dreamed of change, people in search of justice and advocates of a better life. It was one of the finest and most impressive stages in my life.

Where did you find those people?

RMT: They found me! Somehow the people struggling for justice and retribution found each other. They have taught me so much.

Previously, when I left my native country, I had joined a group known as Las Hijas de Maria [Daughters of Maria]. They taught us to operate sewing machines and to clean the convent. My father always took us to church—he was a devout Catholic—where we tended the flowers and other beautiful things. I got on well with the nuns, who taught me to cook and to read and write. Eventually, I worked for a few years at the convent as the cleaning woman. I felt very much at home there. Later on, I found working at private homes to be far more difficult.

Those setbacks are the best learning experiences, in my view. I have often wondered where I would be had I not been through so much sadness in my childhood, if my people and my own family had not been butchered. What would I have accomplished in life? I probably would have become a nun and would never have left my village.

I was at the convent when my brothers were murdered, and I was still there when my father was killed. Later they kidnapped my mother and probably killed her as well.

Rigoberta's voice begins to tremble as she relives these horrible memories. For a moment she seems to want to stop, to put these horrors behind her. Then she continues.

RMT: After the deaths of my brothers, I did my best to help my mother. We tried to continue somehow. Then, when they came to get my father, it all became too much. Our family—or what was left of it—fell apart. At first, I stayed. When that became too dangerous and they came after me as well, the nuns hid me in the convent. Everything happened so fast between September 1979 and April 1980. In those six months, my home life changed permanently, and neighbors, other relatives, and friends of mine were killed as well. Altogether, it permanently changed the course of my life. I was not even twenty-one, and they could easily have clipped my wings and grounded me. But I managed to fly away and received the Nobel Peace Prize at thirty-three. The murder of my parents and family, the slaughter of the Guatemalan population, made me promise myself that I would always struggle for justice for my people. That is a struggle for life.

Since I have dedicated myself to the struggle, I have been delighted to find myself surrounded by people with the same sense of solidarity. I have managed to help others, especially women, who were running from someone or something and needed shelter. This work took me to places of refuge in Mexico and other countries. Being able to help others gives me the energy to continue.

Has the Nobel Peace Prize changed your life?

RMT: Completely! Very much for the better, in some ways. It has broadened my horizons, and I feel I have made excellent use of the ten years that have elapsed since the award. We have launched so many different projects through the foundation, including some with short-term results and others that have been effective over the longer term. In this respect, I feel very successful and regard myself as the most fortunate and privileged woman in the world. In a way, it makes up for all the pain of the past.

Do you ever experience hatred festering within you?

RMT: No, up to now, I have not. I focus on the good things in life and feel protected by so many different people around me. I feel the support of the Mayan people and from all indigenous peoples in this world. Thanks to them, I am never truly alone. But if you are asking whether the Nobel Prize has forced me to make sacrifices, I would have to say it has. In my private life, I am no longer as free as I once was. I can no longer go to the market or run little errands. If I do, people approach me from all directions to tell me their problems, hoping I can help them with money, medication, a new hospital.

In addition to all the wonderful opportunities that the Nobel Prize has offered, I am frustrated that I cannot help everyone. It is difficult not to feel the problems of people asking me for help, but I cannot solve all those problems.

With pain in her heart and tears in her eyes, Rigoberta tells me about a woman from her neighborhood in Guatemala. This woman, like her husband, was dying of AIDS. She decided to leave her three children with Rigoberta. In the end, after the parents died, Rigoberta and her husband, Ángel, found these children a good home.

Then there are the stories of family members of those who disappeared during the junta period in Guatemala. Rigoberta spends a great deal of time and effort tracing as yet undiscovered mass graves throughout the country. In addition to helping others and expressing solidarity, she hopes ultimately to locate the remains of her own loved ones through DNA technology and to give them a worthy burial. This is of the utmost importance in the Mayan tradition and brings peace to the remaining family.

Another great sacrifice that Rigoberta has had to make to travel around the world most of the year is that she can experience only fragments of her own son's childhood.

RMT: The first time I had to go back on the road was only a few days after my son, Mash, was born. When I returned, he was a month old. The second time I returned, he was a few months old. Now he is nearly grown. At nine, he is far more independent and is growing up with his cousins. Fortunately, his father is around a lot. He lives with his father more than he lives with me.

The pressure of all those questions, requests, and expectations from her fellow sufferers clearly remains a heavy burden for Rigoberta. But she knows she is not alone.

RMT: The foundation has been incredibly supportive as I cope with this responsibility. I am truly blessed with a wonderful group of people who work for the foundation all over the world. They each have their own past and relation to the atrocities in Guatemala and elsewhere. This makes us so harmonious and brings us so close to people. Without their dedication, I would be unable to perform this task.

Listening to you, I am moved that you live with such deep love. I see your gentleness clearly—despite everything that has happened. It seems as if fear has had little chance with you; you have not submitted to it.

RMT: I am indeed fearless and feel no sense of reluctance. In fact, if I were to encounter the people who tortured my mother tomorrow, I would give them a chance—first, to show remorse; second, to admit their crime; third, to help me locate the remains of my family and all those others so that we can give them a decent burial and get some distance from the tragic memories that torture us and our children. Fourth, I would give them a chance to ask our forgiveness. Such a request would not fall on deaf ears.

We always say that people who harbor hatred and vengeance will age quickly. And I hope to stay young as long as possible!

What is love?

RMT: Love is everything. People may be tired or frustrated, but not forever. I always look for a joyful side to life. Now that I am writing fairy tales, I feel like a child again. But it also makes me very sad—as I mentioned earlier—to relive my happy childhood, which was later buried in sadness.

I do not have a great many friends. I have many admirers, but I have little time to maintain personal friendships. I cannot phone friends or visit them regularly. But I love to spend time with the small group of friends I have now. We have now learned to cook Mexican food and enjoy eating together.

Love is everything in life. And my husband, Ángel, really is an angel. He is Guatemalan, too, and is a very modest background presence and a source of support and advice. Without a husband like him at my side, I doubt I would be able to accomplish everything I have. Our love is based not only on the fact that we happen to be married, but also because we truly love each other and take good care of each other. His being Mayan as well helps enormously. Sometimes he feels that I am in danger because of everything I do and the places where I travel. He has our spiritual guides watch over our house and accompany me wherever I go.

At the end of our meeting, Rigoberta raises the subject of dignity, which she believes is important for all of us and should be imparted to our children as well. That would be a wonderful gift for the new generation, she says. I offer that the greatest gift we can give the world is to overcome fear.

RMT: I feel that the great thinkers of our era—the people who can help us most in our development—do not want to think much anymore for fear of telling us the truth about this world. We need to make sure that our ideas are not "abducted" and "pledged." We have so little dignity.

Just as we are about to say good-bye, Rigoberta adds something to her description of love.

RMT: Love also means being coherent. There must be coherence between what we say, what we live, and what we do. "Living with love" means being coherent. We need this ethical code to guide our lives and to ensure that we keep listening to each other.

MATTHIJS G. C. SCHOUTEN

Sometimes I think that separation is the sin of all sins, from which all suffering originates.

Matthijs Schouten's dwelling in the Netherlands consists of a tiny apartment perched above an equally small women's clothing shop in a narrow street in the city of Utrecht. A winding staircase starts at the back of the shop and ends in a study, where a remarkably tall man meets me. His hair is short, his clothes are casual, and he wears heavy shoes, as if he is ready to walk out on the bog at any time. He seems a well-grounded person, with the expressive hands of the dancer he once wanted to be. In the room are one easy chair and two straight chairs. From the window is a view of the five-hundred-year-old tower of the Dom church, whose huge clock shows it is ten in the morning.

While he makes us some tea, I look around the room, filled with many Buddha statues and rows of books: The topics of Islam and nature, Kabbalah and nature, Hinduism and nature, Christianity and nature, and Buddhism and nature all catch my eye. There is some Jung, of course, and ecological studies of all sorts. I see a series about animals, literature and poetry, and probably everything that has been written about Celtic civilization. This versatile man also studied Celtic languages and reads them with some fluency. An accomplished academic, teacher, and expert on nature and ecology, he feels the need to share what he knows and chooses to share with students. But he also wants to learn from them through their questions. I will come to understand how important mutuality is to him.

We sit down; I decline the easy chair because I feel I will be more alert on one of the straight chairs, and I inwardly prepare to start with my questions. Looking around the compact space filled with items representing so many fields of interest, I suppress the urge to ask him, "But, who are you?" I have to trust we will learn the answer through our discussion.

At moments during our conversation, his soul shows so purely through his eyes, as he looks over my head into the open air behind the square of the window, that I know he is sharing what is deepest within him, and it is almost embarrassing to continue watching. I wonder if it is all right for me to see so much and I realize how seldom we allow others to look into our souls through our eyes. I feel grateful for his gift and the openness of our talk, and simply wish the morning will never end. My heart opens to his love for life and I feel a kinship with this man I never met before.

I confess to him that after looking around his apartment, and even more so while reading his recent publication, *De natuur als beeld in religie, filosofie en kunst* [Images of Nature in Religion, Philosophy, and Art], I understood that I could start our conversation from any angle, for he is truly a many-sided man. The book was published in 2001 by one of his employers, Staatsbosbeheer (the National Forest Service of the Netherlands), on the occasion of the service's centenary.

Before I realize it, our conversation has started.

Matthijs G. C. Schouten: I wanted to write a book in which I would try to reproduce manifold and different voices with as little prejudice as possible, mentioning my sources so that all readers would be able to find their way to further study, but without venting my own opinions, stating that this and that is good for nature and such and such is not. I leave that to the readers.

But the sad part is that means your presence is not felt in the book. After I finished reading, I was left with the question: Who is he, really?

MGCS: (Laughing) Well, that is something we can discuss now! I stayed out of the book on purpose. I wanted to stay as invisible as possible. That was very, very difficult at times, and I did not always succeed: My best friends told me after reading the book that it shows that Buddhism is my weak spot!

I am not really surprised; I could hear this "hidden love" speaking from his bookshelves. In this beautifully multifaceted book of his, Schouten's love for the Buddhist philosophy of life is pleasantly and subtly manifest between the lines. It is not a coincidence that the volume ends with a haiku by the Japanese monk and poet Matsuo Bashō:

> *When I look carefully*
> *I see the shepherd's purse blooming*
> *by the hedge.*

Schouten explains that the Japanese verb translated as "to look carefully" means something like "to look in such a way that the shepherd's purse can reveal itself." His eyes reflect his enthusiasm.

MGCS: That is probably the most fascinating of it all. While writing the book, I found that this experience keeps coming back in all different cultures. If you are able to look at nature in such a way that your image about it disappears—so that all your preconceived opinions, thoughts, and ideas are pushed aside—what is will become clear. When no image is projected upon nature, it can reveal itself. People like Bashō, but also Martin Buber and the Christian mystics, were able to let nature reveal itself.

That for me is the ultimate form of communication. Hearing you say all this, I am deeply touched because it is exactly my perception of nature. It is the love that shines through each life-form. And it is extremely moving to get in touch with it.

I wonder next how Schouten became attracted to Buddhism. Without me really asking, he starts explaining.

MGCS: Buddhism became very precious to me on the one hand because of the psychological and philosophical method it offers to become a better human being, but foremost because of the unbounded heart toward every living creature. The Buddha speaks about the boundless heart over and over again. He showed a deep care for the Earth and all that inhabits it.

It is said that when the Buddha touched the Earth on the night of his Enlightenment, it was preceded by a fantastic moment. The Buddha was a threat to Mara, the lord of the Earth, or the lord of evil, just as Jesus was. He [Mara] sent his troops to try to corrupt the Buddha and he even put his seductive daughters into action. But the Buddha did not surrender. It made Mara furious; he shouted at the Buddha, "How dare you search for a way by which people can free themselves from their desires and delusions?" The Buddha did not answer; he just touched the Earth with his right hand, because that is where his whole journey toward Buddhahood took place.

Schouten talks about his belief that almost everything evolves from one very important rule: not to see ourselves as separate.

MGCS: Sometimes I think that separation is the sin of all sins, from which all suffering originates.

I have always been deeply touched by these words of the Buddha:

> *Whatever living beings there are,*
> *weak or strong, large or small,*
> *seen or unseen, living far or near,*
> *born or yet to be born,*
> *may all beings be happy.*
>
> *Just like a mother*
> *would protect with her life her only child,*
> *so one should cultivate an unbounded heart*
> *towards all beings.*

That, in my opinion, is ultimate non-separation.

This requires something of us—namely, that we cleanse ourselves so that we can become pure in body and mind in order to be able to look without prejudice, in all openness.

MGCS: Yes! Openness is what it is about, really.

I think that there are two ways to come into being. There are two elementary forces at work. The first one is a sort of centripetal force that causes us to appropriate things. Everything we observe is immediately labeled and then appropriated. The second force is centrifugal; it is the force by which the heart opens itself to the outside world, while nothing is appropriated and space is being given and created. This is the force by which love and compassion are generated. To be able to love, we need to concentrate on this latter force. The former will work any way.

Do we need the first kind of force to survive in this world?

MGCS: (Pausing before speaking) It looks like we need that force more and more to survive, I'm afraid.

For example, here on the streets in Utrecht, I notice that no one seems to be aware anymore of other people in the street; everyone is too busy to get to his next destination. For me, one of the biggest changes in the world is that when I was younger, people did seem to be aware of the fact that they were surrounded by others, which also meant that they would give way to each other, for example. Nowadays, I have the feeling that you almost have to slalom in a world where people are running as fast as they can from A to B, disconnected. One has to take care not to be knocked over. So it sometimes seems that we need that centripetal force out of self-defense. But, then, the Buddha stated in one of his discourses that by protecting yourself, you are protecting the other, and by protecting the other, you are protecting yourself. You should not see yourself and the other as two separate beings. Protecting life—nature included—means stimulating the love within you.

The centripetal force isolates and makes beings into segregated entities. That I would call a downright sin. It means that the access to real happiness is shut off for the other party too.

Matthijs Schouten grew up in the Catholic south of the Netherlands, in a small Limburg village of barely five hundred souls. At a young age he learned by heart all categories of sins, taught and drilled by the parish priest. In his early religious education, the emphasis was on the "infections of the soul." Only later in life did he hear about the God of love. But even in these early years, so full of threats of punishment and fear of sins, he never feared God.

He tells us about one of the first experiences that helped him gain confidence.

MGCS: I must have been five or six years old. I remember it was at sunset and someone I knew well had just died.

I was wandering the fields at the edge of the village, a hazy, red glow slowly covering the fields. All at once, I felt lifted up and enveloped in "being," knowing that everything was going to be all right. For just a very brief moment I had been able to forget about myself.

After that, it seemed as if the concept of God was not relevant anymore, although of course I had to go back to those classes on religion.

Often, deep emotions that occur in our youth will mark us for the rest of our lives. Another incident when he was young may have marked the beginning of Schouten's affection for Buddhism. He remembers how—at maybe four or five years of age—he witnessed a most unpleasant event.

MGCS: One of the most difficult things about growing up in a farming village was the knowledge that all the animals that surrounded me, from cows to pigs, were going to die.

One afternoon, some friends and I wanted to play on one of the farms. Just as we arrived, a pig was about to be shot. The butcher had his rifle aimed. What impressed me most was that the pig did not run away. It just sat there on its hind legs and waited.

As young as I was, it struck me that the pig somehow realized that there was no escape. After it was shot, it fell on its side; the butcher cut the carotid artery and collected the blood. I had to throw up, went home, and was sick for a week!

Something started there. I did not understand that things like this could happen, that human beings could act in such a way and that, apparently, it was even something normal.

When I was ten years old, I read the book *Siddhartha,* by Hermann Hesse. The images that emerge from that story were immediately accessible to me. They showed that every life-form needs space and that no one has the right to take that away from anyone or anything. I guess my interest in Buddhism started from nature, rather than from the philosophical side.

A Tibetan clay tablet on one of the mantelpieces in his apartment reminds him of one of his favorite Buddhist stories. Schouten's voice deepens as he begins.

MGCS: I have always been so impressed by the story of the deity with eleven heads and one hundred arms, who in India is called Avalokiteśvara. The story tells us that at a certain moment this being was ready to enter final Nirvana. He had practiced during countless lifetimes and he had developed enough to gain full enlightenment. But on the very moment that he was going to realize Nirvana, he suddenly heard a scream of pain. Looking up, he saw a hare caught in a trap. He was so upset by the suffering of this animal that for a brief moment he was dissuaded from his resolve to make the transition to Nirvana. All of a sudden, he saw suffering everywhere and he knew then that he could not go, for he felt unable to leave behind the suffering creatures. That is when he grew eleven heads, to survey all directions in order to see who needed his help, and one hundred arms to reach out.

(Short silence) I think this is exactly what it is all about. The minute that selfishness, greed, and aversion stop, there will be *metta,* as it is called in Buddhist terminology. It means to give space, to let be, so that everything can be what it is—which also implies that the suffering and joy of the other can and will be fully experienced.

As I again cast a quick glance at the clock on the Dom tower, I realize that we have been talking for a long time, though it feels as though our encounter has just begun. Schouten has so many interests—among them biology, philosophy, and Buddhism—that I want him to speak about all of them.

Do you consider yourself to be a pioneer, not in the least because of your rare combination of talents and professions?

MGCS: No, I see myself more as a connector. In my earlier days I wanted to discover or invent something, to become famous. Now I believe my role in life lies in trying to create space, dialogue, communication.

I was able to resist the criticism of the people around me who warned me against not making choices in life because, so they said, I would never succeed in anything. They tried to make me feel that I was acting selfishly by focusing too much on my own world, exploring my divergent interests, instead of being productive in one field. Now that I teach, I finally see how everything I have learned and developed is useful—and how I can share what I know with my students. I think that if I did not teach in university, my life would definitely be less fulfilled.

We need a little break and, while Schouten makes us a second cup of tea, I remind myself that I want him to reflect further on the role nature has played and is still playing in his life, for it seems to find a place in all his interests. To begin with, nature had a strong hand in deciding to which field of science he would dedicate his time.

MGCS: As a young boy, my father created a small garden for me in the nearby hazel woods. Not yet ten years old, I was sure that when I grew up I wanted to work with plants. I stayed loyal to gardening and my father's garden for more than ten years, but ultimately decided to become an ecologist in my professional life.

Do you feel that people are part of nature?

MGCS: The very fact that I am a biologist makes me say yes. The Earth is our very own ecosystem, which we need to survive. I don't agree with many philosophers, most of all the ancient Greeks, who thought that because humans have self-awareness and reasoning, they cannot be seen as part of nature.

The Romantic German philosopher Friedrich von Schelling saw nature as an organism that, in a self-creating process, gradually develops her own possibilities. Starting from minerals, she has unfolded herself through ever-more-complex forms of life, ultimately—in the human species—to realize her self-awareness.

So where does that leave us? Given that everything on Earth has a task, then what is our role in nature? We Westerners talk about "conquering mountains" and "taming nature." We always seem to be the conquerors, while in Eastern countries, people talk about an encounter with nature. This must be a difficult question for someone who is both a philosopher and a biologist at the same time.

MGCS: (Carefully and thoughtfully) As a biologist, I should say that our main task is to make sure that we do not exhaust the Earth and that we do not exploit and plunder her. It is a tragedy that a species with the extended brain capacity that we have has been able to develop so many instruments of destruction! From a biologist's point of view, I sometimes think of us humans as a frightening species. As a species, we joined in on the very last second of the evolutionary process. Then, just think about how much we have destroyed only in the last century, a period that is no more than a nanosecond in the evolution of life! There are nights that all of this almost keeps me from falling asleep.

What is our task? Do we have a task?

MGCS: As a biologist, I sometimes really don't know. But from a spiritual point of view, I see the enormous challenge that our "being" is putting to us. On the one hand, we have the capacity to express our greed and anger in an unrestrained manner. We possess all the tools to do so, and if we don't, we will invent them tomorrow. But while we possess this "magic wand" that could make our biggest dreams come true but may also realize our deepest nightmares, nature on the other hand is constantly and consistently telling us: No! Stop!—because there is a world full of life, a world full of other beings who also need space to be able to survive. So, whoever wants to give space to the fullness of life has to conquer his or her greed, anger, egoism, and so forth.

I'm curious what Buddhism has to say about our place in nature.

MGCS: Characteristic of Buddhism is the concept of continuity of life in the sense that no clear distinction is made between different life-forms. Human beings are seen as exceptional, to a certain extent, because of specific capacities they possess. But in the Buddhist view, there is no sharp division between humans and other beings. During the training on the Buddhist path—which is more than anything a training in letting go of self-centeredness and in developing compassion—we are taught to focus not only on our fellow humans but instead on all that lives.

I remember that Schouten cites a striking anecdote in his book Images of Nature *that illustrates this. As the story goes, mice were invading a nunnery in Tibet. The nuns caught the animals and released them outside the monastic compound. But the mice came back in ever greater numbers. Eventually, there were so many mice that the nuns could not walk anymore without the danger of crushing an animal. They decided to leave the buildings to the mice, then, and set up a new nunnery elsewhere.*

I liked that story.

MGCS: (Laughing) My own house was once filled with mice too. My first reaction was fear—not of the mice themselves, but of what they were going to do to my books! Then I realized that there was not much I could do, so I just let them be. One night I came home and found a dead mouse on the floor in the middle of my study. The strange thing was that I really felt a little sad. I buried the creature in my roof garden. After that I got used to their presence. But when I came back from a trip to Ireland last year, they were all gone. I have never seen them again.

Apparently they were there to show you something.

MGCS: Without a doubt, they showed me something. My feelings on mice and the way I see them have completely changed this past year.

I tell him about my experience with the red-winged monarch butterfly and the message I felt it came to deliver, and Schouten reacts immediately:

MGCS: This is fascinating because it sometimes strikes me that we consider it normal and accepted to go to a concert to find inspiration. But society's reaction is somewhat more skeptical when you say that you need a long walk through the wild boglands for inspiration. For me, to see the first bog asphodel come into flower on a gray and overcast day in May in Ireland is a wonderful experience; in that encounter I feel everything else slowly fade away and it brings me great inspiration. That is when nature is really telling me something.

Is it difficult for you sometimes to combine your status as a Western scientist with being a Buddhist?

MGCS: I do not feel a field of tension. As a scientist, I know that our rational way of approaching things is just one way to look at reality: a very important one nevertheless, since it has given us so much. But as soon as scientists state that it is the only way to look at reality, I feel that they are excluding a lot—namely, everything beyond the rational concept.

I guess you could say I am a scientist educated in the West but one who considers science not to provide the final truth. In the Buddhist view, ultimate reality, ultimate truth, lies beyond rational thought, beyond any concept.

It was Schouten's biology side that was attracted and seduced when he started work in the bogs of Ireland. At first he went there to study bog vegetation. He started a comprehensive survey of plant communities and their environmental requirements. Years later, he still goes back to the Irish peatlands, but his journeys have become part of a bigger effort. The Irish bogland has turned out to hold the perfect appeal to him as both a biologist and a Buddhist.

MGCS: The blanket bogs in western Ireland cover thousands and thousands of acres. They are treeless, so you never feel protected anywhere.

Walking the bog meant that, in a certain way, I became very aware of my physical being and of my own vulnerability, literally hearing my own heartbeat because of the silence that surrounded me. If anything, those bogs have taught me to be modest and to open myself.

It was really in an almost meditative way that I became aware of myself and at the same time, in the vastness of the peatlands, could completely forget about myself, because nature simply *was!* And then, after a while, I just *was* as well. And then we could *be* together.

I had learned that as a boy, Schouten had taken ballet lessons. He seems surprised that I know this fact.

MGCS: It had a similar effect on my life as the bogs had years later. It was an adequate counterpart to the education I received at the Episcopal college, which was focused on solely the mind and the soul.

To my teachers, the body was sinful and no more than the vehicle for the soul. But more important, being in motion while dancing was the way to forget about myself, to not think at all. But my body became too big and I could not do what I saw myself doing in my mind's eye: rising above myself, extending, becoming light, weightless. So that was the end of that.

What began as a survey for his dissertation grew into something larger. After the first year of fieldwork in the bogs, he came back the next spring to continue his study and found that a number of bogs he had visited the year before were, in a manner of speaking, bleeding to death.

MGCS: They were using big digging machines to drain the bogs and to extract the peat. I felt that a system that had existed for over ten thousand years was dying before my own eyes.

In order to stop the "killing," he started a campaign to save this part of nature. His battle brought him to the United States to make a plea to the U.S. Irish community and to Brussels to seek the help of the European Community. He also tried to show the Irish people what they were losing. After ten years, the work paid off: Irish politicians decided to create a net-

work of peatland reserves so that this vulnerable ecosystem would survive.

The bogs obviously touched you. What was it that really made you take up and continue the fight to save them?

MGCS: Lots of people have asked me that question and I was always tempted to answer: Someone had to do it! But looking back, it feels as if I was being asked to do this by the bogs themselves. And whenever I go back now, it is like visiting my friends, like going home.

It is a home that includes for him an associate professorship at both the University of Cork and the University of Galway.

It is evident that Schouten is one of the people who experience an "interbeing" of all life. He agrees.

MGCS: It is a beautiful approach to being. Nothing can exist on its own; everything is dependent on something else, which means that there should always be openness and a way of communication based on a form of mutuality.

(Citing the haiku poet Bashō) "Go to the pine if you want to learn about the pine or to the bamboo if you want to learn about the bamboo. And in doing so, you must leave your subjective preoccupation with yourself. Otherwise you impose yourself on the object and do not learn."

This is the way in which you allow nature to reveal itself. You could call that a form of communication, which in my view is very important in our relationship to nature. For me, communication means not separating or appropriating, and this always implies mutuality. And mutuality begins with openness, real openness, in which, as Bashō said, you leave behind your subjective preoccupation with yourself. Then the other—whoever or whatever—can reveal its being. And we must be prepared to take in what is shown, even if it does not suit us. We must be prepared to change our views of ourselves by what is revealed. In that moment, mutuality comes in and communication begins. Too easily we think that we are open to reality when we are really just listening to what confirms our self-image. This selective openness can never lead to mutuality. I would even consider it as disrespectful toward the other.

I reflect on how "selective openness" in this last case has an almost paternalistic aspect, which indicates a consideration of ourselves as better than the other. This leads to dominance, whether of man over woman or of people over animals and other forms of nature. Separation again. Within communication, things can be so much more pleasant! The cheerfulness in nature is as strong as the ruthlessness. There is definitely great humor, lightness, and playfulness there.

The bouquet of flowers on the table between us is an example of this. They stand there, radiating health and beauty. He refers to a friend of his who always says that God put in a great deal of wit when she created the flowers.

Does Schouten have a view on where we stand with the world? Some have said that the world will end in the year 2012. This frightened me until I came to think that because we will all die at some point, we would do best to live in love and in the now rather than in fear of what might happen. At the same time, I am appalled by the immense field of tension we have created with nature—a tension that we can't ignore any longer.

MGCS: Absolutely, and we have done so for a long time . . . In the West, we have taken nature for granted. We have used her, cultivated her, and only rarely have we treated her carefully. What is happening now is a worldwide development that goes beyond cultures, making it fascinating and frightening at the same time. The ecological problems that we are facing, created by the almost violent way in which we have come to claim nature, not only affect her, but will also have consequences for us.

Because of the worldwide dimension [of the problem], we are no longer able to hide behind science in the hope that it will find us a solution for environmental problems. We can no longer say "It will be solved in the future," because that future is so close that we are forced to act now.

If you ask me where we stand, I would say that in general we are probably more aware of the problems on a global scale. We communicate to each other about the need to care for nature on a worldwide level, which means that we have to learn how to interpret each other's images, words, and experiences. All these developments will, I hope, lead to the understanding that we as the human race can no longer claim more space. Ecological problems force us to think about who we really are and what we are doing here.

There is a spiritual side to it, but what do we do in practice?

MGCS: You cannot separate the spiritual from the practical. Maybe in this case you could say that technically speaking, the practical part is how you manage the conservation and survival of ecosystems and species, while the spiritual part has to do with the questions "What sort of being am I?" and "Why should I care for nature?"

Do you think it matters if species become extinct?

MGCS: Yes, to me it does, because every species is part of what we are. That is, by the way, the same kind of question that I always ask my students. I do not want to give them my answers, but I do want to teach them about communication, because I am convinced that asking things is the best way to hear things. It is important to listen to what you do not yet know.

I want to know if you assume that all life-forms have awareness.

MGCS: (As if struggling to find the answer) I think that if we define *awareness* as we define our own awareness and self-awareness, it will not apply to all life-forms. But, of course, you will find plenty of awareness in nature, on all different levels. (Pause) Well, I realize your question makes me feel as though I have to find common ground among the philosopher, the scientist, and the Buddhist.

Your encounter with bog asphodel moved you, as you told us earlier, which could mean that there is awareness there.

MGCS: (Even more puzzled) That is my restraint: I can only experience my part of that encounter. The other side, I cannot.

***Would you prefer to talk about* soul *instead of* awareness?**

MGCS: No, soul is even a more difficult concept for me. In the Buddhist view, there is no soul, because nothing is permanent.
But I guess it is simpler. While writing the book, I found a text in the Midrash that says: "When a tree is cut down, there is this almost inaudible cry that pervades the whole world." I remember that I was really touched by those words. Now I think that the fact that I was so touched probably means that I have heard that cry. End of story!

We have come near the end of this morning's inspiring conversation. One question remains—one that is relevant because it plays an important role in the relationship between humans and nature.

What is love?

MGCS: (After some silence) A Buddhist master once said that his enlightenment started when he suddenly saw his own face in every face he met, and every other face in his own. That for him was love. Of course he meant *face* in the broadest sense. I think that is what it is. For me, *love* is a tricky word, because there it can have many interpretations that have little to do with love, that are more concerned with appropriating, desire, and fear of losing. But if every face is visible in your face, and your face in every other face, love can never be the "wrong" kind. You are where the sin of separation is absent, and there will automatically be compassion for those who suffer and sympathetic joy with all those who are happy.

Works by Matthijs G. C. Schouten

"Disturbance, Conservation, and Restoration of Ecological Systems: Dreams and Realities." In *Disturbance and Recovery of Ecological Systems,* edited by P. S. Giller and A. A. Myers. Dublin: Royal Irish Academy, 1996.

The Image of Nature: A Cultural-Historical Perspective. Utrecht, the Netherlands: KNNV, 2005.

"Nature Conservation in Ireland: A Critical Decade." In *Biology and Environment* 94B, no.1 (1994): 91-95.

"Nature Conservation in the Netherlands." In *Educating for Environmental Awareness,* edited by J. Feehan. Dublin: University College, 1997.

Peatlands, Economy, and Conservation (Matthijs Schouten and M. J. Nooren, eds.). The Hague: SPB Academic Publishing, 1990.

GARETH PATTERSON

The health of the planet and our own inner health are one.

Knysna, South Africa. I wake at six, excited to meet the "lion man of Africa." This early in the morning, the air smells fresh and sparkling. I grab a few grapes and a bottle of water and head for the Land Cruiser. I always love the little climb from the footboard into this mighty car, and the deep sound of its diesel engine as I turn the key and start on my way into the mountains behind Knysna. I drive through the rolling hills that form a beautiful far-stretching rhythm; tree plantations cover part of them. The greens of the land are soft and soothing to the eye.

Finally, a winding dirt road leads me into the forest and to some houses on the edge of the wood where Gareth and his partner, Fransje, live. I find them in the last house in a row of wooden houses, from which they emerge, ready for the walk we had decided to take. They are preceded by several happily barking dogs of different shapes and sizes, who somehow manage to shake their whole bodies while wagging their tails, no doubt hoping to join us on our forest adventure. I had an image of Gareth from his picture on the back of his book *To Walk with Lions,* but seeing him now, he looks different, older perhaps. His hair is longer and he is taller than I imagined. He seems to be very much in his own world. I detect an inner silence, an inner voice to which he listens. Right now his face still shows the signs of a deep sleep.

Fransje is lovely. A bouncing mop of light brown hair encircles her smooth face. She wears immaculate white gym shoes. How will they emerge from our walk? I wonder.

The three of us drive deeper into the huge forest, which grows denser as we penetrate the area where Gareth is actually searching for three remaining, and by now legendary, elephants. He has found signs of their bodies rubbing a tree and has noticed uprooted plants and droppings. They are out there somewhere, swallowed by the forest. They could easily be standing silently only a few yards away from where we are, hiding behind some scrub, and we would never detect them. One was seen for the first time only last month. Gareth has named her Strange Foot, because of her strange spoor, and he gathers from it that she must be about sixteen years old. One of the others is an adult that had been seen for the first time a year ago after a disappearance of twenty years. And, then, there is a younger one Gareth recognizes from its small spoor. He believes there might be others that have never been seen.

Why has the man who adopted the name "the lion man of Africa" switched to elephants? And how has he coped with the killing of his great lion friend, Batian? I have many questions on my mind. But first we begin this walk to get to know each other.

We don our rucksacks and head into the wilderness. A narrow footpath cuts through the lush, humid grandeur of the Knysna forest. Strong, huge ferns tower above our heads. I enjoy the humidity of the air and deeply inhale its strong scents, while already feeling the warmth of the sun starting to seep through the dense growth. We hardly talk, but connect through a love for the wild forest. It is beautiful, and the fact that we just might encounter "the mighty animal"—the elephant—hidden in this wilderness makes our walk exciting and adventurous.

Gareth merges with the surroundings, walks with his senses wide open, sniffing, touching, and pointing out the smallest detail. Although we are not lucky enough to find the elephants this morning, we do feel their energy around us. It is their forest.

Back at the car, he rests his lean body against the vehicle and talks about his work and ideas. I am fascinated, watching as his face seems to change into that of a lion; his long hair frames his leonine features.

In his book *To Walk with Lions,* Gareth introduces himself as "a man who has lived among lions" and as a "lion man" who has lived among modern people:

> Living in these two worlds, I have seen "wholeness" in lions and I have seen "disconnected-ness" in people. Man, Nature and God have become separated in the modern era, leaving us spiritually and emotionally impoverished. When we see ourselves as part of the natural world, destruction of the natural world can be recognized as self-destruction. By acknowledging this as a fact, we can begin to heal ourselves and the natural world.

We drive back to Gareth's and Fransje's home on the last few drops of diesel—apparently I was so excited this morning that I forgot to refuel.

Fransje has made sandwiches in the early hours of the morning and, with a delicious cup of South African rooibos tea, we start our conversation.

What is the message that you think lions are passing on to humankind?

Gareth Patterson: Lions have always been present. All through history and in all cultures, you see lions as symbols of strength and power. The image of the lion is used in heraldry everywhere, representing courage, power, and strength. I would add, as did George Adamson, that they represent protection, healing, wisdom, and serenity.

Jesus was called the Lion of Judah. So was Haile Selassie, as the hypothetical returned Christ. He always walked with two lions next to him. King Solomon was at times symbolized as a lion because of his wisdom. In certain parts of Africa, the lion is associated with God. In Egypt, lions were kept as sacred animals. In the temple of Amun-Ra at Heliopolis, lions were cared for by the highest-ranking priests. At the death of a lion, it was decreed by law to have a full public mourning. They carved motifs of lions to decorate their beds, believing that humans had to be protected while asleep at night. These symbolized the Aker lions, the double lion gods of yesterday and tomorrow that protected afterlife. The Aker sphinx lions were closely identified with the enigmatic Akeru, the Lion People, who were said to have been the gods on Earth in pre-Dynastic Egypt.

Don't forget the leonine symbol of all time: the Sphinx! Some researchers are suggesting that the Sphinx at Gaza dates back as far as 10,500 B.C.E. and that when it was first created it directly faced the constellation of Leo just before sunrise.

(Short silence) The Basotho and the Tswana of southern Africa use the word *tau* for the lion, which means "the star creature" or "the creature that came from the stars."

In Latin America, the Aztec and Mayan shamans had the power to become half jaguar and half human. In western Africa and the Congo basin, the phenomenon of the leopard man was well known and may persist today.

I want to tell you the story of a man called Ed Flattery, who farmed in Botswana. Two of his cattle were killed by lions on two different occasions. Furious, Flattery set off with his Bushmen employees to follow the killers' tracks, determined to catch the lions and stop the killing of his cattle. The spoor led them beyond the farmland into the Kalahari Desert. Finally, [the trackers] saw smoke rising in the distance and the lion tracks led straight toward this same place. At this point the Bushmen got nervous and refused to proceed, explaining that it was not the lions that they were following but the Makaukau, members of a Bushman clan who are known for their ability to transform into lions.

They told Flattery that they thought these Makaukau had transformed themselves into lions to hunt during the night. Flattery persuaded the Bushmen trackers to proceed, because he wanted to find out more about this. The tracks led them to two small grass huts. The smoke they had detected came from a small fire burning in front of one of them. The lion tracks continued toward a large pile of ash in the center of the cleared area. The tracks crossed the ash and then simply disappeared. No further tracks could be found. There were two men and two women sitting near the huts, with their children. Both men had injuries, one in the head and one in the chest. The Bushmen employees were convinced these injuries had been caused in the battle, by the horns of the cattle.

Gareth gives me a long look. He has saved his own lion experiences for last.

GP: I have had some odd things happen myself. Several times, I have been somehow "on the move" with my lions through the bush at night, while at the same time I was fast asleep in my tent next to my girlfriend! Certain game rangers in the Tuli bushlands reported that, on some mornings, they found my tracks and those of the lions at sunrise. These rangers are superb trackers, and by reading the spoor they were convinced I had been walking with the lions a few hours previously, in the total darkness of the night.

A book about the Bushmen's art, called The World of Man and the World of Spirit: An Interpretation of the Linton Rock Paintings,* *mentions how one Kalahari shaman claimed that a man in lion form is able to mix with a pride of lions without fear. A Kung shaman uses the words* a pawed creature *to describe a man who travels out of body in the shape of a lion. To nineteenth-century Bushmen, so powerful was a man in leonine form that they believed a lion that did not die after it was shot must be a shaman.*

So you would not be the only one, Gareth. You could be a lion shaman.

GP: (After a silent period with his head down) At first I just thought it impossible and laughed when the rangers told me about my tracks. But then, one early morning, I came across my own spoor with the pride of lions, and the spoor was fresh. I had not been in that part of the park for quite some time. And that same night I had not left the tent. I can't explain this. I only know that on both occasions, I felt a bit tired in the morning. (Looking at Fransje) I haven't told you this, but it's happening again, and I have the same feeling of tiredness in the morning.

The connection—call it a spiritual connection with my lions—ran through a deep, almost subconscious level. In part, I was living for them, and I think I would have died for them. Yes, I was like a father

---

*J. D. Lewis-Williams, *The World of Man and the World of Spirit: An Interpretation of the Linton Rock Paintings* (Cape Town, South Africa: Ed. SA Museum, 1988).

to them, but it ran even deeper than that. They represented every lion that ever lived and represented every lion that humans have destroyed. It's beginning to feel a bit like that with the elephants, too, now.

It may interest you to know that other people—Sai Baba is one of them—have reported being at two places at the same time in their own body.

GP: (Looking at the soles of his well-worn shoes, a bit lost) Yes, I was in my body, my feet made imprints on the earth. I saw them, so the weight of my body was there, making the imprint. I was definitely in my body when I made those spoor.

This opens a new set of questions: Where were the lions? Is everything an illusion? Are we living parallel lives in which we exist in more than one place at the same time? Is it through bonding with the soul of the lion that Gareth can go beyond what we call earthly reality, as the old cultures tell us?

GP: The soul bonding existed between the lions and me. When soul or spirit reunites for a reason, it can take place beyond the human-created "species barrier." After all, we all have souls, regardless of our physical bodies. But how it actually works, I don't know. It has happened a few times. And each time, I could recognize the imprint of my boots from tracks only a few hours old.

There is another odd thing [that has] happened to me, something I can't explain either. I was in the United Kingdom with my mother, where we went to a wildlife park to see the rare descendants of the Barbary lion, extinct in the wild for almost a century. As we got to the lion enclosure, I decided to hold back until a large group of people had moved on. I wanted to connect mentally with the lions, without other people being near me. The lions could be distracted by the crowd. While my mother and I watched, we saw how the lioness—born in captivity—and her cubs ignored the people. As the group moved on, we approached. The lioness immediately became enraged when she noticed me, fully charging me and stopping just short of the enclosure fence. I was surprised by her behavior and moved back, leaving my mother where she was standing. The lioness continued to snarl as long as she could see me. As I hid behind a tree and other people moved to where my mother stood, the lioness once again ignored the people. But whenever I came out of my hiding place, she started charging me as before. I had not even started talking to her!

For me, the answer to the repeated behavior of the lioness is not a logical one. It is instead to be found on the metaphysical level. After spending much intense time among the lions at the Kora National Park, the lion aspect of my nature would have been very strong and apparent to her. She was clearly defending her cubs from a threatening, non-pride male lion. She was not reacting to my physical presence, but rather to the spiritual aspect of me.

What do you mean by your "spiritual aspect"?

GP: It is my very core, my soul's connection to everything around me. It is also an inner voice that I think we are all able to hear but, to varying degrees, don't want to listen to.

***How do you understand* spirituality?**

GP: For some of us, it is the connection to what we see as the divine, personal godhead, for example. Spirituality is connection regardless to what faith one believes in. In conventional religion, the connection can be by prayer. In order to live a day-to-day spiritual life, I feel that we have to be aware of our own connection to everything around us, so we can feel empathy for others, including all forms of life.

What happens when you realize that there is so much more to life than what is accessible through the five senses?

GP: It helps me overcome the trauma and the sorrow about what happened to Batian, my beloved lion son. He was lured out of the reserve and shot dead by trophy hunters.

The following excerpt is from Gareth's book To Walk with Lions:

> On the second anniversary of Batian's death I walked to where I had buried him, beneath a cairn of beautiful stones. I had taken his iden-

tification collar (which I had put on him for a while) down to the grave and placed it in the hole of a tree near the cairn of stones. Now I put my hand into this hole, expecting to find the collar. To my surprise I found it was empty. I then thought that perhaps an elephant with its curious, reaching trunk had found the collar in the hole, pulled it out and dropped it somewhere nearby. I searched thoroughly in the nearby bush but did not find it.

I stood beside the grave before moving away to sit beneath the tree in which I had put the collar. Sitting there I began to visualise what Batian would have looked like had he still been alive. He would have been five years old; he had been big for his age and I believe he would have grown into one of the largest lions ever to range the Tuli bushlands. As I sat there, I willed that I could receive some sign from him. After a while I stood and walked back towards the cairn of stones.

Suddenly, upon the dusty ground, I saw the fresh paw prints of a lion. I had walked on that same piece of ground earlier but had seen no tracks. It is impossible that I could have missed them. In the bush one reads the ground like a newspaper and one's life can depend on what it tells one. The size of the paw prints were the same as Batian's before he was killed.

I took a big breath in an attempt to calm myself and recover from the shock of seeing the tracks, then began to follow where the paw prints led. Incredibly, the tracks led me directly to the cairn of stones and then onwards. I continued to follow the tracks—then suddenly saw Batian's identification collar directly where the paw prints were leading. I had received my sign from him.

The last time I had seen the collar, it had been buckled to form a circle. But that morning, when I found it, it lay stretched out upon the ground. It had not been unbuckled but somehow had been cut. I picked up the collar, held it to my chest and spoke to Batian. Today, I firmly believe that Batian was showing me that he lives on in spirit. By leading me to the cut collar he was leading me to the understanding that I should make a break from my grief (hence the collar being cut and laid straight out upon the ground). To this day, the experience is very comforting to me. Every day I know that my beloved Batian is never too far away from me.

In all that lives around us our loved ones who have passed over watch us, care for us and I believe can present signposts in our lives to help guide us on the path we walk on Earth.

It is an incredible consolation that I know that Batian is with me. I don't know, but I wonder when lionesses come and sit with me and appear to see something that I can't see, or when lions come and walk behind me to greet something there, if it isn't Batian they feel or see.

One evening before she returned to her cubs, Furaha, Batian's sister, walked with me down to the grave. There, as I sat on one side of the grave, Furaha sat on the other side. The sun lowered with a growing blaze on the western horizon. It was unusually quiet and Furaha and I strongly sensed Batian's presence around us. Beyond a streambed an impala herd crossed in front of us. They did not see Furaha or me. We watched them pass, and later we rose and headed back to the camp in the golden light.

The trauma of Batian's death has been so severe for Gareth that his body has since taken to slight trembling. At times it gets worse. He tells me that, in fact, he is a very shy man. Noise, riding in a speeding car, or traveling in a busy city is intolerable. In the company of large groups of people he is soon exhausted.

Do you think that when a person becomes more sensitive and learns to be in contact with and open to all life-forms, such a person will have difficulty living in society?

GP: No. I have drawn upon Batian's courage whenever I stand in front of an audience to present a lecture on my life with the lions. I draw upon it whenever I am confronted with adversity or criticism, or when I received death threats after exposing the whole "canned lion"* business.

What is the message the lions give us?

GP: (More quietly) I think the healing power of the lion is going to be very important. Did you read the last page of my book *To Walk with*

*This refers to the shooting of lions in an enclosed, confined space.

Lions—"A Final Word"? It is so important to acknowledge the healing power of the lions. Like the lions, we can let the healing come from within.

On this last page, Gareth mentions attending a meeting to raise funds for a haven for adults with cerebral palsy. A woman approaches him:

> She told me that around the time that my first book was published her son was involved in an accident that left him in a coma. For days her son was read to and spoken to in an attempt to elicit a response from him. But this was to no avail. The lady began reading her son stories about the Tuli lions from my book *Cry for the Lions*. His condition improved remarkably—and when he came out of the coma the very first words her son uttered were, "What happened to the lions?"

I believe our personal healing is linked to the healing of the Earth, and this is one powerful way we each can contribute to reverse the destruction of the Earth. It means that every person on Earth is important, which I find beautiful. Do you agree with this?

GP: I have come to understand that unless we address the health of the Earth, collectively and holistically, the symptoms of our own inner health will persist and will worsen. The health of the planet and our own inner health are one. The more we took from the Earth, the more impoverished we became. Nothing is meant to be separate from the whole.

Gareth was born in the United Kingdom and was brought to Africa, where he was raised, at the age of eight months.

Did you have any experience in your childhood that called you to the lions?

GP: At the age of six I was in a jeep full of grown-ups. My mother was there. We went to the bush and I saw a lion. But nobody else had seen him and they did not believe me. Later the lions were found; I had sensed them before the others could see them.

My mother made me an awesome plaster model of a game park

with mountains and rivers and animals, the lot. Then I knew. (Laughing) At age eight, I applied for a job as a game warden; I had seen it advertised. I got a very nice letter back explaining that the place had been taken already.

I went to school in the United Kingdom, where I visited the local zoo. I saw the lions there imprisoned and in awful circumstances. It is the starving of the soul that is the worst imprisonment. I remembered the image I had from the free lions I had spotted in the game park, and the enormous difference has left me with a taste for freedom ever since.

Freedom in my heart is essential to me, and I think it is for all creatures. I was in the United Kingdom for my O or A levels, but my heart was in Africa. I became out of balance and got an eating disorder. I would force myself to vomit after eating too much, mostly sweets and stuff like that. On holiday, I talked about it with my Malawian friends, who were some ten years older than I. They suggested traditional medicine and I agreed. During a small ceremony, they inserted a concoction of herbs into both my shoulders. This medicine became a part of me. When I returned to the United Kingdom, I felt better, stronger. The medicine was protecting me against jokes and insults because of my being different. My confidence grew slowly and I realized I was transforming myself by believing in myself.

Did you ever feel that you had to get a job and go to university and all that?

GP: No, I knew my heart was in Africa.

So actually you always remained who you are?

GP: Yes, I did. And there was the help of my African friends. At that time, I sat down to write a letter to the old man George Adamson, who lived with the lions, telling him about my love for the wild and how I wanted to learn from him.

While listening to Gareth, I see a strong yet vulnerable person—a sensitive man with inner strength who knows who he is, someone who has no need to pretend in the way that so many of us do . . . those of us who fol-

low the path we think the world expects or demands we travel, and thus become slaves to our desire to make "the right impression."

I also see a man who now has the strength to remain vulnerable. He walks his way through life, open to where his intuition tells him to be, for this is where he will be, even if he sometimes doesn't understand why. He knows that in time, he will learn the reason. He speaks as if he can read what I am thinking.

GP: I don't yet know why I am here with the elephants, but I know I have to be here.

Did George Adamson reply?

GP: His wife, Joy, answered me. But because at the time I was still finishing my school, the letter had traveled around quite a bit. And then, on the very day I received her letter inviting me to come and work with her, she was murdered by an ex-employee, who stabbed her to death. So we never met.

But Gareth did start working with her husband, George Adamson.

GP: My career in wildlife started at the bottom, working in a game reserve in South Africa, carrying the tourists' luggage. I ended up as a field guide in the Tuli bushlands in Botswana. There I became fully involved with the lions. I made a friend named Darky—a male lion with a magnificent black mane. Darky accepted my presence, and I regard him as my lion father. Lions in the parks are justifiably fearful of men, and his acceptance of me was therefore extraordinary. He taught me much about his sort. Fifty percent of the Tuli lion population died by the hand of men in not more than two and a half years' time. It was in Tuli that I learned a lot about the threats to the lions in Africa. I wrote a book to draw attention to the dangers to the lions and made a twenty-five-thousand kilometer journey of discovery through southern Africa. I wanted to learn about the history, the present status, and the possible future of the lion in that region. After that, I felt I was ready to write to George Adamson, telling him about my study in Tuli. He invited me to come.

It was very special to meet George: finally, someone who shared my love and concern for the African lion. At the end of my visit, he asked me to join him in his work with the orphan lions. I returned within three months and worked with George for half a year. George's wild pride consisted of second- and third-generation descendants of lions that he had rehabilitated in the past. At the time, he had adopted three small orphan lion cubs, which he named Batian, the male, and Rafiki and Furaha, his sisters. At the age of eighty-two, George was dedicated to ensuring that one day the cubs would find their freedom back in the wild. He loved those little ones, and it reminded me of the lioness Elsa's story. He had telepathic communication with the lions, as if the lions and George had two-way radio contact through their hearts and could speak to each other. I learned a lot from this memorable man. There are only a few people who have had the great privilege to share lion fellowship. It is a unity of souls. Once you attain it, you will radiate it.

Toward the end of my stay in Kora, George asked me to work full time and, after me, to make sure that the work in Kora would continue. But I felt I had to be with the lions in Tuli, although I felt tremendously honored to be asked to work with George. I left. On my way to Nairobi, through the open window, a dove crashed and died against my chest. It was a sign of things to come.

On August 20, 1989, George was gunned down, killed by poachers. I heard the future of the orphan cubs he had been raising was insecure. I could not imagine them in captivity, but did not wish to be in Kora myself. The authorities closed the whole area for security, so the cubs' rehabilitation could not continue there. It was simply too dangerous. I then asked Richard Leakey to help me transport the lion cubs to the Tuli bushlands, because I wanted to continue George's work in rehabilitating the lions back into the wilds. The plan succeeded.

The following is from To Walk with Lions:

> Rafiki is, as I write, twelve years old. She is a grandmother, and she roams in her territory in the Tuli bushlands with her pride, her family of eight. Through all the hardships, the tragedies and the despair we shared, Rafiki and her pride represent to me triumph over adversity.

> Through her and her offspring and their offspring, other lions will continue to be born free—and so the Adamson lions live on.

Have you ever been afraid with lions?

GP: No. When I was with lions, I felt a great calmness within. I saw the same thing with Adamson, who was always very calm whenever he was with the lions. It's a place inside that knows.

I once came back to the Tuli camp after a longer absence. I walked into the bush and met my lioness Rafiki; I embraced her, putting my arms around her. Looking up I gazed into the eyes of her new mate: a wild lion! We looked at each other and I felt this peace within. He looked at me in equal calmness, totally relaxed. It was all right.

Did the lion know you?

GP: He might have. The weird thing is that this lion looked exactly like Batian.

Was this lion as old as Batian when he died? Could it have been a reincarnation?

GP: It did cross my mind. But the time frame did not match. Nelion (Rafiki's mate) would have been alive the same time Batian was alive. They were a similar age. It did seem, though, that Nelion was connected to Batian, a soul brother perhaps. His arrival and the healing consequences of his arrival were not a coincidence, that's for sure. Rafiki and I needed his arrival. She and I, the last survivors of a shattered pride, needed the intervention from a higher source. That is when Nelion arrived.

That the lions and Gareth are deeply bonded is obvious, as this excerpt shows:

> One evening my small pride appeared at my camp after having been away in the eastern portion of their territory for several days. As usual, they greeted me affectionately. During our fond greeting, I suddenly noticed that a large fluid-filled membrane hung from Rafiki's

vulva. I became concerned, as I feared that she was in the early stages of miscarrying cubs. The following morning, I set out in search for her—like any concerned parent would. Despite tracking her for several hours, I was unsuccessful.

That afternoon I recommenced my search and quite suddenly came across Rafiki, alone, not far from the camp. It was as if she simply materialised in front of me. Immediately, I realised there was no sign of the membrane or of any bleeding. In her own special way of calling and skipping, she indicated clearly that she wanted me to follow her. Puzzled I did follow her for a while, but as the sun began to set, I had to stop and turn back towards the camp.

Early the following morning, I found Rafiki beside the camp's fence. Once again, she repeated her behavior of the previous day. I stepped out of the camp and with Batian at my side, began following her through the bush. We followed her for over an hour and during that time she repeatedly turned around, called to us, and then continued to lead the way. She led us up to a rise and then entered a particularly thick clump of bush beside a small crevice. I saw her climb down into a thicket and then, while hidden, heard her calling us. Batian stepped forward, and ventured to where she was. I watched as he peered downwards into the thicket. He remained there, unmoving, for several minutes, then moved backwards and lay down.

As he moved aside, I stepped forward. I looked into the thicket and saw Rafiki nestled in the bushes with a perfectly formed, though dead, cub lying between her paws. I was amazed. She had been so determined to lead us to see her cub. I felt a strange mixture of emotions: great sadness for Rafiki that the cub had been stillborn, and at the same time touched and enormously privileged that she had so wanted Batian and me to see the little one.

Gareth lived with the lions for almost four years. As he says in his book:

It was almost a surreal life. I hunted and played, rested and defended territory with my small pride. On one occasion the lions saved my life from an attack of an enraged leopard and upon the birth of cubs both lionesses led me to their newborns. I was part of the pride and I saw

> life through lion's eyes. This experience, living as a human member of a lion pride, allowed me the privilege of entering a dimension of lion life that one would not have thought possible. The essence of the relationship between my lions and me was based on mutual empathy, trust and love.

Tell me about the moment you realized the oneness of all life through Batian.

GP: I walked out into the wild with Batian as I felt he was ready to give his first roar. We were facing the rising sun and I had my right hand on his flank as his deep call reverberated across the land. Time stopped and I felt I was that roar, I was the lion and the lion was me. I saw myself in everything: I was the Earth and the Earth was me. I was free.

Did this change your life?

GP: I realized my true belonging to all life around me. It was a moment of re-connection, feeling connected to our ultimate mother: the Earth.

Later, I came to realize that the "connecting energy" is essential to access if we are to free ourselves from the modern illness of loneliness of spirit and the sense of no purpose. We are now at a point when we know, even unconsciously, that we have to reconnect. Our survival on this planet could depend upon it. It is time we reconnect spiritually with all natural things. We have lost the understanding of "I am because we are, and because we are, I therefore am."

The white man's religious belief—unlike indigenous beliefs—did not allow him to feel part of the environment, but instead led him away from it. The only connection was to use the Earth for selfish reasons. There was and is no reciprocity, [so] characteristic of tribal societies. But, who can really persecute you for loving the Earth, for loving yourself, for understanding that humans are part of the web of life, and that all has a purpose? What are we afraid of?

Also in the city, you can reconnect. Healing comes from within. If we attend to the emotional affliction, we can heal the physical. The notion of healing is a wisdom well known in the African life. In the West, we have come to separate the spiritual from the physical, treating the symptoms, not the cause. The African believes: Treat the whole person, body and mind. By treating the emotional cause, healing follows.

Lions have the most incredible recuperative powers. Humans, I believe, can draw inspirational powers from the lion.

I feel the unconditional love in nature and it is obvious to me that by feeling the oneness of life, you touch that love.

GP: Yes, unconditional love. It is a love that I can describe only as being like a light reflecting from a beautiful crystal into all directions—a light that reflects to and from our souls, a love that I believe transcends what is referred to as death and reflects into the spheres beyond earthly reality. I have seen it in the love between the lions and George. My lions have taught me about it and opened me up to it.

How can Gareth combine feeling this unconditional love with being an activist, which most often originates from anger and outrage? For example, how did he combine it with fighting the "canned lion" industry, as he calls it—with waging his intense campaign against breeding lions for the gun and against the disgusting trophy hunting linked to it in South Africa, as described in his book Dying to Be Free?

GP: I felt it was something I had to do. In the presence of lions, I feel whole. I have an insight to their ways and can feel their hurt. This hurt is my reason for trying to speak out for their kind. To me, the shooting of lions for sport is simply abhorrent, morally wrong. As a simple murder in man's society deeply affects the whole family, so it is with lions. I believe it is as wrong to murder a lion as it is to murder a man. Trophy hunting occurs because there are those—the client hunters—who will pay money to justify an urge, a yearning for what they perceive to be a masculine performance.

"Canned lion hunting" can be defined as the hunting of usually captive-bred lions in confined, fenced areas. The size of the area can be a cage, an acre, twenty hectares [fifty acres], up to one thousand hectares [2,480 acres]. These different areas have in common that they do not represent a lion pride's true territorial range in the wild. So to shoot a lion within a fenced area that is not accepted to be a natural, territorial range to me is a canned hunt.

I wrote to a newspaper in 1990: "The story of the lions offered to hunters on the 200-hectare [496-acre] farm distresses me greatly. What further levels will man stoop to for his own selfish entertainment?"

Through personally living intimately with the Adamson lions in Botswana and knowing these animals as individuals, recognizing in them a range of emotions akin to our own, I cannot see a barrier that divides humans from animals. In South Africa, only now are we moving toward the realization that all humans are equal. In turn, we must recognize that all creatures are our equals and should be treated as we would expect ourselves to be treated.

Surely, former political activists can agree that there are strong parallels among racism, sexism, and speciesism. Hunting in South Africa is a two-hundred-fifty-million-rand industry. Credo Mutwa, while giving his opinion of the white domination and racial overtones related to conservation and wildlife generally in South Africa, once said to me: "Conservationists and hunters are to me among the greatest racists and biggest bigots I have ever come across."

From its earliest founding, the Kruger National Park was established on land where Africans were forcibly removed from their homes. They were made to live in native reserves or locations beyond the reserve's boundaries. They also lost access to wildlife as a means of subsistence.

Wildlife hunting, or the death game, as a sport and recreation is very much a white male activity in South Africa. It is not to fill the stomach, but rather to satisfy an avaricious yearning for what is frequently perceived as a masculine performance—the ultimate demonstration of power and control. It is an industry that lives on death. It is a white man's conservation ethic that animals have to "pay their way."

I recall to Gareth that Richard Leakey, Kenya's world-renowned anthropologist, conservationist, and politician, came to South Africa in 1997 to debate the question "Does wildlife have to pay to stay"? Leakey said:

> The only way to win this battle is to avoid the price tag. I am not personally opposed to wildlife utilisation. But restricting it to private

> [game] reserves run largely by Caucasians is like sitting on a time bomb that will go "bang." Bio-diversity must not be regarded as the preserve of the foreigner. It is unrealistic to think we will go forward by saying that species must pay to stay. It is *Homo sapiens* who must pay.

GP: We have reduced wilderness to mere islands in comparison with the wilds of the past. Surely, these last places—with the animals that inhabit them—should be allowed to exist solely for their own sake, without the financial demands of humans made under the guise of conservation.

To be able to live in Africa, we must have an understanding of African things: We must lose our Western arrogance and seek an African perspective on the natural world. I am not referring to the Westernized African attitude. I am talking about precolonial African environmentalism, the kind that Credo Mutwa tells us about in his book *Isilwane* [The Animal]: "We believed that human beings could not exist without animals, birds and fishes, or the greenery that whispers around us. We used to believe that in every one of us there lay a spiritual animal, bird and fish with which we should keep contact at all times, to anchor our family upon the shifting surface of this planet."

It is an African belief that wild animals belong not to man but to God. To kill for sport is therefore a crime against God. In contrast, to protect wild animals, you have to respect them, and in order to respect them, you have to feel for them, and by feeling for them, we reflect God within us.

I feel it is no coincidence that the most extreme form of trophy hunting—the callous breeding of wild animals purely for the gun—exists in the country where a white, Western minority clung onto undemocratic rule for the longest period in Africa. The utilitarian conservation was born and promoted by the same ideology of the old apartheid doctrine, in which control and manipulation are vitally important. Trophy and sport hunting is an unholy relic from the past, and I hope that it will be increasingly recognized as an activity of colonial origin, belonging to that past. The idea, however, that hunting pays for conservation is frequently and loudly touted by the hunting fraternity. Let us not forget that the early settlers' belief was that they had to tame the wilderness—and that their religious convictions held that wilderness was an insult to God, and that the principal of "dominion" should prevail.

I ask Gareth if he also feels that at the moment, we are in a spiritual conflict between two groups. On the one hand, we have those who believe or want to believe that we humans are on this Earth as the highest in rank in the chain of life, free to use the Earth and its inhabitants in whatever way we want. On the other hand, some believe that we humans are an intrinsic part of the whole and that the interconnectedness is holy in itself, meaning that we have to reconsider our place in the web of life. Credo Mutwa would characterize this attitude as the Christ child within being born.

GP: Well, yes. And in the case of the wild animals, the pro-use people see the animals not as beings with a fundamental, natural right to simply exist in areas of suitable habitat. Instead, they see wild animals indeed as possessions to be bought, bred, sold, and hunted for economic gain. The minute the Tuli lions crossed into South African game farm country from Botswana, their death sentence was signed.

I tell Gareth that when I read the book Dying to Be Free, *tears streamed from my eyes. I was profoundly shocked once again by the cruelty humans are capable of committing. I was desperately in need of fresh air; I had to set aside the book and go for a walk. We have moved so appallingly far from the concept of oneness with all life, from respect of life itself. Sometimes it seems as if human awfulness has reached a point beyond repair. Humans are the only cruel animals on Earth: fully aware of what we are doing when we kill and destroy. Those of us who are isolated from oneness, who are cruel, are conscious that we kill and destroy for the killing alone.*

Of course, it is not only in South Africa that humans treat animals with no feeling or emotion, as if they are at our disposal. This happens wherever humans live. In the worst scenarios, this attitude is horribly and tragically applied to other human beings as well. I cite a passage from Dying to Be Free: *"Two men on a drunken hunting trip failed to find any deer and instead cold-bloodedly murdered a deaf black man yesterday, as he walked along a railroad track in Chico." (from the* Los Angeles Times*)*

GP: It seems clear that trophy hunting detrimentally affects entire families, not just the killed animals, in a variety of ways. Of course, it also affects the chain of life. Our understanding of the behavior and social structures of the different wildlife species is slowly increasing. We are recognizing the devastation caused by the killings. The death of so many prime pride males has a shattering effect on the pride, and social chaos spirals. Males are shot, mothers with cubs are shot, siblings are separated from siblings, and lions are broken in spirit: It is clear that this creates imbalance in the whole social and family structure and all life around it. By protecting a lion, you protect all that comes beneath it in the food chain: prey species, trees, plants and grasses, and the wilderness, because the lion needs a vast territory.

If hunting would stop—if ever!—the wild would take generations to recover. I believe that the damage done by trophy hunting has affected all life physically, ecologically, and spiritually. "Expelling God from every day life leaves the field clear for the super capitalist, the colonist and other plunderers to rape the Earth, to destroy nature, to ravage priceless natural resources with cold impunity," says Credo Mutwa.

We are witnessing a total ignorance of the species. The Western notion, rooted in the traditional Christian belief that objects to any reverence shown to nature, has made the exploitation of animals acceptable. It seems that people who regularly harm or kill animals become increasingly desensitized to what they do. What kind of human beings do they become? How are their relationships with other humans affected? And how do they relate to themselves?

Killing lions is made so diabolically easy for the hunters that I reckon even my grandmother would be able to do it, in a manner of speaking. Children are offered to shoot, as are disabled and sick people. They even have special programs for the disabled. The lions are often drugged before they are shot from a vehicle.

I will never forget the dark lioness who was separated from her siblings by a fence and shot from a vehicle in front of her cubs. The scene was filmed by the ranger, who was unable to stop the carnage and had the courage to expose the killing through the video film. We got to see how the dark lioness twisted through the air just as the bullet hit her, in a last attempt to be with her cubs. The report shows scenes of the milk from her teats mingling with her blood on the floor of the skinning shed. She had still been feeding her cubs.

I wonder if the international and local hunters really believe they are hunting in the vast wilderness, or whether they actually know they are killing in partitioned and fenced-in "wild" areas. Just remember that the word used for wild animals is *game*—the fun sport of killing the animal!

As I researched the murky world of hunting, the most repugnant cases of hunting and killing that I learned of increasingly began to haunt me. It affected me deeply and now that I knew about it, I wanted to make it all public. Animals have never deserved the way man has treated them. We should make an effort to continue to lift the lid, not only on sports hunting but on all "profitable" institutionalized cruelty to animals.

When I lived among the lions, I saw for myself the parallels between their emotions and those of humans. I have been with happy lions and I have been among the anguish of a lioness when she realized that her sister and her two cubs were dead.

Listening to Gareth, my heart is bleeding. How do we reach people who have this destructive perspective? With books such as Gareth's or the one you are holding? With movies about the wild? By banning violence from television? As Rigoberta Menchú points out, it makes no sense to punish those who no longer have access to their feelings. The key to transformation lies rather in them opening up to feeling again. Arne Naess brought criminals into nature to help them return to feeling and seeing and understanding that they are part of it. Here lies the means to transformation: When we are able to love ourselves better, we can love others more easily. We have grown away from ourselves and from the connection to all life. We must remember that animals have their own right to be and live in a way that is natural to them. We need to remember that animals, plants, trees, and water—indeed, all life—is not ours to take.

It is ultimately a matter of awareness and education. We should start with children, explaining to them the falseness of our Western idea that humans are the center of the Earth and the universe. Those who speak in this book point out that loving nature means studying nature. In getting to know more about different species on the Earth, it will occur to us that animals, too, have feelings, emotions, a soul, family structures—life.

GP: I believe that the pro-use and the anti-use arguments will continue to exist until there is a spiritual reconciliation between man and animal. This will occur only when we dare to understand, admit, and feel what we are doing to the animals. Only then can a spiritual reconciliation, empathy, and a renewed respect be born.

A bunch of wonderful people made a documentary on the killings [the canned-lion hunts] titled *The Cook Report,* aired in the United Kingdom in 1997. Finally, it reached South Africa and caused an enormous uproar, internationally as well as nationally. The International Fund for Animal Welfare (IFAW) office received fifty-five thousand petitions of protest against breeding for the gun industry here.

I want to quote Dr. McCallum, a psychiatrist. He wrote an article in the *Cape Times* entitled "Man's Inhumanity to the Young—Adults Should Listen to Children to Recover Their Humanity."

> Ask children what upsets or distresses them most in the world and they will tell you. Their answers are consistent; it is in their speech and in their art: The sad and frightening things that people do to each other. The sad things that people do to the animals and the trees.
>
> Ask the children what we, as adults, should do about it and they will tell you very clearly: "Tell them to stop it."

I wrote my book *Dying to Be Free* in an attempt to reach as many people as possible in order to stop this. I—as did all those who made *The Cook Report*—received death threats, and we were under a lot of pressure to keep quiet. Now that it has become an international issue, at last there is a greater awareness. The human conscience has been stirred, and in some cases there has been a reaction from politicians.

(Softly) To come back to your question about how I combine the love I feel for nature with my anger, there came a moment when I had to cleanse myself of the anger and the hatred I felt inside and had encountered. I could have clung to my great anger; it would have given me an opportunity to be heard. Sometimes, we stick to our anger as an identity, as something that empowers us. But I felt that great anger enslaved me. It sapped me emotionally and spiritually. I understood the lessons to be learned from the pain I felt within; I learned that by hating, I was not healing. Once I let go of it, I could see that people's hateful actions can be born from great harm done to them sometime in their life. Today, I visualize the light of goodness upon the person who caused my hurt. This might help that person to release his or her pain. Being angry does not help the other in any way.

How did you release your hatred?

GP: I decided to go back to Kora, George Adamson's last home. It was quite something to be walking there. I went there with a longtime friend of George. The strangest thing happened: All George's pets, including the ravens, came to greet me. I walked onto the high Kora rock and let go of much of the darkness the lion killing had brought me. I felt Adamson's presence in all life around me. He was there, in every leaf, in every tree.

(Whispering) It is quite something if a person becomes All. He was really there, in all living beings.

Gareth is open to all the energy of life around him, to all these influences.

GP: Now you can understand why I have moved to the forests of Knysna to research elephants. It was too much to bear, for too long. With the death of apartheid, so too must die apartheid in the field of conservation. With the African renaissance, we need to seek and develop the environmentalism of the continent. I am not a conservationist; I am an environmentalist. Conservation has too many ugly things attached to it. I too believe a change in attitude has begun, undeniably linked to the death of apartheid and the birth of a democratic process in South Africa.

As we come to the end of our conversation, I ask if there is anything he wants to say before we close.

GP: Yes. I strongly believe that the lions can heal people and I think more and more people are getting in contact with that energy. The understanding that we can interrelate with nature, with Earth, is beginning to return today into many people's minds.

As I mentioned before, Credo Mutwa calls this the Christ energy that is returning, the Christ child being born in all of us.

GP: There is such wisdom in the old ways. It can be found in a single grass seed, in the wind beneath an eagle's wings, on the rough, intricate bark of a tree. The Earth speaks to us. We can listen to the Earth and by doing so we can begin to understand. The Earth, men and women, the air and the water, and the sun and the animals: Everything is one holy One. The health of the planet and our inner health are one.

Gareth is a man in contact with his inner voice and, through the lions, with all life. Isn't that the same as being in contact with God? Is God not manifested in all life-forms, in the lions as well as the grass and the wind? And when we live in obedience to that inner voice, are we not joining ourselves with the Creator to create? I believe Gareth, as well as Jane Goodall—and I have met many others—is a channel of the animal world, which is showing us the way to renew our connection.

GP: (Looking at the ground) Every step we set on Earth should be conscious. We should be conscious of every step that connects us with Earth. We should be grounded in who we are. We can reach out again and reconnect.

Works by Gareth Patterson

Cry for the Lions. Sandton, South Africa: Frandsen Publishers, 1988.
Dying to Be Free. New York: Viking Penguin, 1998.
Last of the Free. London: Hodder and Stoughton, 1994.
The Lions' Legacy. London: Robson Books, 1991.
Making a Killing: The South African Canned Lion Scandal. Preston, England: The Captive Animal Protection Society, 1999.
To Walk with Lions: The Seven Steps of True Spiritual Fulfillment That Come from Living with the King of Animals. London: Rider, 2001.
Where the Lion Walk. New York: Viking Penguin, 1991.
With My Soul Amongst Lions. London: Hodder and Stoughton, 1995.

Web site: www.garethpatterson.com

CREDO VUSAMAZULU MUTWA

If we could say the word love,
the whole universe would dance.

South Africa is a place on Earth that I find highly impressive and challenging in many ways. It has stunningly beautiful nature, with enormous contrasts in landscape: from the deserts in the north to the lush humid greenery in the tropical east, the flat-topped mountains that rise out of its plains, its wild coasts and huge canyons and well-tended farmland. I run a nature reserve in the mountains of the Karoo. Sometimes the days are blazingly hot, other times bitterly cold. Winds howl, storms rage. Sand is blown around, dust settling everywhere on this dry land. But when, occasionally, the rivers are full after blissful rain falls out of those huge black clouds, everything sings. Then, waterfalls provide us with lovely back massages and the animals find sweet green sprouts to nourish themselves. This semidesert is a powerful country and I feel deeply connected to this part of the Earth, which I experience as strong and abundantly giving. South Africa, I find, is a part of Earth where both the cruelest and most creative sides of people can come forward. Here I have met many remarkable people.

The great skies, the thundering and sometimes frightening storms, the canyons and majestic views from the tops of the mountains, opening up endless shades of grays and blues—these all together shape the people here and give me strength to be who I know I am and to do what I want to do.

Meeting the Inyanga, also called Credo Vusamazulu Mutwa, the Zulu *sanusi,** was exciting. There are only a few sanusi in South Africa. They heal nations and individuals. They carry a vast knowledge and, therefore, a huge responsibility. We might compare Credo's specific wisdom of the Earth to that of the Babylonian wisdom of the skies. I had come to know him through his impressive books, in which he reveals himself as a storyteller of great and abundant skill, and I ardently wished to meet this man who can cite Shakespeare as well as old tales of the Zulu people. His life is dedicated to healing the sickness in the minds and bodies of people in his country, the sickness between white and black, as well as that of the many thousands here who suffer from AIDS. As a selected initiate of the Bantu (African people of South Africa) sacred Great Belief, he broke a vow of secrecy, accepting all the consequences of this decision: He has willingly revealed this old and profound knowledge and wisdom for all to know today through the stories of the Bantu people.

Credo believes we should all gain access to this higher knowledge offering us a choice to avert the catastrophes he foresees. He is a man of great controversy, not only because of this courageous step to make available once secret knowledge, but also because he refused to take sides in the South African struggle with apartheid. His own life has been threatened several times, his son was burned, and his loved one, his wife, was killed. Many of his family members have died of AIDS.

I had asked various people their opinion of him and, interestingly enough, all had different views. One thing they all said was that he is a formidable man. I learned that he receives few people and that I was lucky to talk to him.

I go to see Credo with Helen, a friend, and we find him behind heavily guarded fences in a small and modest house. Once again, I am struck

*Credo says a *sanusi* is a healer of the Old Class, a healer of nations. He or she worships God as the Mother and must serve people, with a duty to be an "uplifter," to create jobs for people. A *sangoma* is a healer of the later time and belongs to the later man. He or she is served by people and worships God as the Man.

that the wisest among us live so simply. Among his many talents, he is an artist. A huge metal statue, one of his constructions, stands tall in the grass. It looks like a robot or a figure from *The Wizard of Oz*.

A small white dog, shaved to the skin, comes running out, barking a high-pitched challenge. The bodyguard comes to greet us and ushers us through a small kitchen to a bare, nondescript room without soul that obviously serves to receive visitors. "Please sit down," our guide tells us.

Into the room walks a big man. His enormous body is decorated with a chain of jade representing a snake. Hanging from the chain is a large rose quartz crystal embedded in copper, under which dangles a huge copper ankh, the sign of life. The whole chain must weigh more than is usually comfortable to wear. He is covered with a sky blue woolen wrap fastened by a large pin under his chin. His attire reminds me of pictures of my grandmother Queen Wilhelmina during World War II, when she too wore a blanket around her shoulders, being in her way as modest as Credo appears to be. His eyes shine behind thick glasses, his head tilts slightly to one side, and he wears a smile on his face.

I ask him if he is willing to participate in this book.

Credo Vusamazulu Mutwa: Your ladyship, I wish what you are doing could be done all over the world, by all thinking people. Many years ago, the world became very happy, when people thought that Communism had collapsed, when the Berlin wall was taken down and there was no longer an Eastern and a Western Europe. We were not, because whenever you remove one big danger, you often open the door for two or three other big dangers. And this is what happened. As we are talking now, millions of people are dying throughout Africa. (Holding up his hands) In these hands I have held men and women, some of whom died. When you are a high traditional healer, you must take your patients into your hands and try to instill some of your life force into these patients. I have done that so many times that I am sick of it. Africa is being destroyed without being given a chance.

But Credo, as I understand it, the Great Belief that you live from has much to offer people in the whole world. The notion of the oneness

of life, the knowledge that humankind is an integral part of the whole of nature is so needed at this moment—as is your profound knowledge of medicinal plants.

CVM: Yes! But we are being stopped.

My experience is that in all of us, there is a more or less conscious knowledge—maybe I should say a feeling—waking up that we are part of a larger whole; that what happens to one person on Earth happens to us all; that the wars out there are the wars inside each of us; that we are part of nature and that by destroying her, we actually destroy ourselves . . .

The muscular white bodyguard interrupts here to say that they have offered a video series documenting the sacred knowledge of the Great Belief to the minister of education, free of charge. The minister's answer: We are not interested. Credo's people offered it to the government as well, and they were not interested either. The guard explains that the people at the grassroots level want to hear it—there is such a hunger for this knowledge—but the powers are not letting it out. I interject:

All the more reason to write a book like this.

CVM: But what is the use of a book, Your Highness? You see, Africa is being deliberately, cold-bloodedly destroyed, and some of the destroyers are the very people who are the children of Africa, because in order to destroy a people, you must activate some of the people from within the community to [attack] their own heritage, their own culture. If you look at Africa, you see strange things that are totally out of character to the black man everywhere. There are wars that are tearing up many African countries—wars that begin nowhere and end nowhere. At the end, each country's content has been so utterly destroyed that it can't be called a nation anymore. It has happened right here in South Africa. The country of Lesotho has been totally destroyed. Did you know that the people from Lesotho came from a land called Ntswana-tsatsi? Did you know that the people from Lesotho knew the ancient Egyptians? Ntswana-tsatsi is their name for Egypt and means "land of the little sun hawk."

The colonizer knew exactly how great the black man was, but the knowledge was kept hidden from the normal people, all the better to dominate us and destroy us. Wars, hunger, and now disease are the result.

Africa has a tremendous thing to give to the world, ma'am, but such is the mind of the African that he takes everything for granted. The African is not surprised when he is able to find water underground. Virginia here, this initiate of mine, she was trained by her father how to find water underground. There are hundreds of women like her in the drier parts of South Africa, but nobody knows about them. You know about all the great psychics in America, but no one knows about the thousands of South African psychics who die unknown. They are contemptuously called witch doctors.

Credo, do you think that, with a little bit of time, we will all be open again to that knowledge?

CVM: Your Highness, there is no time, ma'am. When a whole continent of two hundred and fifty million is being destroyed, there is no time. Who are you going to learn from? When we are swept away, when people like Virginia are gone, who is going to tell the story of Africa? It is a story that says: Over there is a tree. That tree, Your Highness, do not see it as a tree. It is I. The African is taught not to say that a tree is a tree. No, it is I. Now, that is very hard to understand. If that tree rots away, I must find out why. Because it is I who am dying. When the white men came to this land, they found Africa full of animals. There were millions of springboks, of elephants, of buffalo, and they thought: Well, this is how it is. They did not know that we were preserving these animals because we saw them as ourselves. Some of the cruelest laws that our people had were laws for preserving animals. If you killed a leopard without telling the king, the killer and his entire family were executed. If you cut down a tree like that one standing there in front of us, the moshabele tree, without telling the king why you wanted to cut it down, your stomach was cut open, your intestines were pulled out and were tied to a tree identical to that one, until you died. If a boy was caught urinating into a river, a piece of goatskin was tied around his penis and he was tied to a tree so that the skin dried and closed up his organ. He could not urinate for a whole day, and sometimes he died.

I was brought up in a mission school, and I was taught that our people, ma'am, were savages who knew nothing. The missionaries told us this again and again. Did you know that in my time, the missionaries forbade marriage between Christian blacks and non-Christian blacks? Like gods, they talked. My mother and my father were separated because my father impregnated my mother, which was a terrible crime in those days. Sex before marriage was strictly forbidden. My father, a Christian, had lost his wife in 1918 and was looking for another wife. He found this Zulu girl who resembled his dead wife and wanted to marry her. She was a non-Christian and such a relationship was forbidden. To try to keep the girl, my father made her pregnant in order to appeal to the missionaries to allow them to marry. They refused. You can't marry a heathen, they said, and the two were separated. I grew up despising my own people, thinking they were heathens. The black people who wore beads, skin skirts, leather lion skins—we used to call them Satan's people, stinking people. As they said, they were working for King George in England; we believed them. Then one day, Your Highness, there occurred to me something that brought me to my senses.

(Taking a deep breath and looking at his hands through his thick spectacles) I used to work for my family, collecting bottles, which I still do. I cleaned the bottles and painted pictures on them. I sold these bottles to mine workers, to send to their wives. Then one day, a group of mine workers from Mozambique chased me and raped me. It happens to many young men in the mines. Suddenly, my world was turned upside down. Nobody wanted to know about me. My father chased me out of his home. I was alone. But because my family needed my money—it was in the days of depression—I was asked to leave my earned money outside the house; I was not allowed in. It was considered my fault that those men raped me. I tried to appeal to the missionary fathers, but they did not want to know me anymore. They did not help me when I needed it most. The Christian church closed its doors to me and left me utterly confused. I found that the white man's religion was a lie.

Then my half sister told me that the woman who had brought me up was not my real mother and that she could take me to Zululand, where my real mother lived. I was half mad by that time and we rode on the train and arrived in Zululand.

After traveling long distances on foot, I reached my grandfather's village. I did not know who these people were. I was afraid because I had been told they were Satan's people. All of them wore skins. Women were smeared with red clay, their breasts hanging out—everything the missionaries hated. I did not dare accept my mother, a heathen who smelled of herbs and who wore beads and was smeared from head to foot with red ocher. But deep in my childhood, I remembered this woman.

Was she happy to see you?

CVM: She was. But she repelled me. It took me all that I had to accept that old man with one eye, a terrible scar on one shoulder, smelling of animal skins, as the father of my mother, and that unknown woman as my mother. It was very, very strange. But these people in whom I had not believed managed to do for me what white doctors could not: They saved me. I was told that there was another religion. And when my grandfather opened my mind to this, he told me how the history in the history books was all a pack of dirty lies. There was so much the books could not explain.

Can you remember what it did to you as a little boy, getting in touch with this new thinking, getting in touch with nature?

CVM: I suddenly felt . . . how can I describe this? . . . fulfilled. I felt complete, Your Highness. There had been a part of me that had been missing. My eyes were opened. I asked my grandfather why I now felt strong, after he gave me bitter things to drink, and why my mind felt so calm. He returned my questions, using a very rude word: "Because you have drunk your Mother." I thought he was insulting me, but he was telling me that at last I had come back to the Earth. I am a man with many faults, but I have one thing: To me, knowledge is everything. I have the need to search for it.

(Showing me the picture of an older lady on the shelf next to where he is seated) This is my aunt, who became my teacher and told me I had a choice: I could live among the gods of my people or die at the feet of the white man's God. "Choose!" she said. I had thought my people worshipped demons, and now I learned they were God.

Can you remember your emotions?

CVM: First, I was scared and felt guilty. I found that the missionaries and their books had been lying to us all along. It was confusing.

Guilty of what, Credo?

CVM: I was breaking away from the white man's God. I was afraid of Satan. Now I wanted to know and know more.

I found that our people in Africa knew how to make cloth; they grew the best cotton in the whole world. But, then, why did they not wear cloth? Why not skirts? The cloth was only for menstruating, but not to wear. The animal skin was to wear, so as not to scare the animals. The hippopotamus does not kill people, as is said in the nature reserves. They did not kill our people, who used to go into a river that was full of hippopotamus and gather water there. My mother—who was one of the leading women—used to go and speak softly to the animals, saying: "Hippo, hippo, we come with respect to fetch water," and they would walk in single file because the animals are not threatened by people moving in a long line. They feel threatened when we approach in a semicircle and when people don't smell of nature anymore. The Zulu women smelled of sweat, herbs, animal skins, Your Highness. They smelled of nature.

Today, people smell of soap and perfume, and the colors they wear are very offensive to the animals in the wild. If you fight against nature, then nature will react. But if you are with nature, it will be with you. Living in harmony with nature is not a small thing. Our people had very cruel laws in order to keep Africa alive.

My tribal totems are two: One is the crab, who walks sideways; the other is a zebra, from my father's side. The green mamba, the leopard, and the lion are from my mother's side. We are not allowed to even think of killing a lion, for by killing the lion, I would be killing myself. Every tribe had an animal that they needed to honor and could not kill. If we had to kill an animal that was possessed by an evil spirit, we had to seek out somebody from a tribe with a different animal totem to kill the dangerous one.

We believe that humans are not the leading animals on this planet. We believe that the animals were created before us. We believe that humans are here to serve the animals by protecting them. We are not here to kill them.

Do we serve them by protecting them or by living together with them?

CVM: By living together, ma'am.

That's very different from protecting them.

CVM: Yes, but look at it this way. There is a severe drought. Where are we going to find food? We lived by hunting, ma'am. But we were not allowed to hunt the pregnant animals. You were not allowed to hunt more than one buck in ten days. Exactly what was needed? There were very strict rules.

It was not that everyone knew just how much was needed?

CVM: No, ma'am, the tribal chief told you when to hunt and when not to. If you would go to Zululand now, to where the modern way has not come yet and you could see a Zulu girl fetch water, you would really laugh. You know what she does? She looks in the water to see if there are any little insects. She will take them out. She must not kill these insects.

His words makes me think of Arne Naess and how he will not step on anything with life if it is not necessary to do so. His whole philosophy of deep ecology is based on this respect. I ask Credo if he has heard of deep ecology, but he is unfamiliar with it.

How was it for you to live in Western culture later in your life, knowing all this, having the old knowledge within you?

CVM: I was like a prisoner. I saw my people going mad. I saw my people's families falling apart.

And how was it for you personally?

CVM: Very bad. I was like a creature with two heads, one in Western society, one in African society. The white people taught us that nature is dead, but my mother, my aunt, and my grandmother used to tell me that nature is alive. They used to say: "You are alive, are you not? Have you ever seen a dead thing giving birth to a living thing?"

The mother of my mother, even when she was about to die, took me outside and asked me to look up into the stars as she said: "Son of my daughter, those stars are alive. Out there are things that are alive. Those stars gave birth to the Earth, which is alive, and the Earth gave birth to us, who are alive."

Do you know, Your Highness, that our people used to do many strange things? Do you know that when Africans dug a mine, they needed copper that was dug by women and melted by women, especially by those who today are called lesbians? They were the sacred women who were empowered by ancient law to work with this metal—they, and men who could not make children. They were very holy people to us . . . also bisexual people. They had special jobs in our communities. Gay people are now persecuted. For example, Robert Mugabe says gay people should be arrested. But his people, the Mashona, used to respect gay people, who were sacred to the Mother Goddess. Some of the wise people who taught me what I know were gay people. Everybody was respected, had a place.

Living in the West, among people who are not connected to nature, did that at times make you start to question your own belief?

CVM: (Very softly) No, ma'am, because when everything has been taken away from you, when your freedom has been taken away, what is left? Nature! Nature will never let you down.

I sigh because I know this well.

Yes. It seems nearly ridiculous to ask you this, but how do you define nature?

CVM: Something I need and that needs me, that is the African way, if you know nature, ma'am, and you really love it.

My eldest son was dying of AIDS about six months ago. We went to help him and he became strong again, but his friends rejected him. Today in South Africa we see a terrible thing happening, which is the result of people cutting themselves off from nature and humanity. In the old days, when someone had a bad disease, all would gather around him to give support and herbs. Now AIDS patients are hated, rejected, and also killed.

When you are taken away from nature and see trees as trees and no longer as a part of yourself, you lose that power to feel. How can I put it? When you see an animal, ma'am, with one leg injured and you have a look, you want to do something for that animal; you want to help it—because, as you said, it is you. If you allow that animal to die, you are killing yourself in the future. That animal could maybe give birth to many other animals, which could give you food or help you live in another way.

While I was in the West, I lived in a strange society where people feel with their brains. But a person who lives with nature, ma'am, does not feel with his head; he feels with a second part of his body—(pointing to his abdomen) down here. It is a feeling a mother or father has when something threatens his or her baby. That is the nature feeling, a feeling deep inside. We must return to that feeling—a complete feeling of being one with everything else.

Also, my lady, I will keep on referring to that tree. Please have a look at it, because you will see many such trees all over this place. That tree can change the history of my people, even now. Just as sutherlandia* has done. And that karee tree over there is used to make the breasts very big.

---

*Mutwa here refers to a plant, also called kankerbossie (cancer bush), that is said to enhance our immune system and to help cure illnesses.

The word *ndebele* means "long and large breast."* Taking the berries at the right time made the breasts big. As for the warriors taking the berries when they were ripe enough, they had only to rub the leaves on their bodies to make their muscles strong. They developed on their own, without needing to exercise the muscles. Don't ask me why, but it is true.

(Softly, with head lowered and eyes looking sideways) I will die with this knowledge, which could create work for thousands of my people, because when the world knows about this, the karee tree will be finished in South Africa.

There are trees that are of great value to human life, we know. One of them is a tree the white doctors call very poisonous, but it is not if you use it at a certain time. *Euphorbia* is the Latin name and this tree can cure cancer—really cure it, like kankerbossie or sutherlandia can heal AIDS. What we have known can change the whole world for the good. But our people no longer know. Our people die here of AIDS, and yet kankerbossie grows along this river. It grows in the wild.

No one has the right knowledge anymore. When men were building a bridge down here, they cut down many of these plants because they thought they were just rubbish. When Virginia and Sam, another initiate, went down there and asked to have the stuff that was cut down, the man said they could take "this bloody shit." One of those men had just told Virginia he had AIDS. He said he had not long to live; he was getting weaker and weaker. Virginia told him he was stupid, that he was cutting down the exact tree that could save him! He started drinking it, and today he is driving a tractor like nobody's business!

A terrible crime was committed when our people were cut away from nature, being forced to work in the mines. The white man brought ideas with him that were foreign to our people. In old Africa, ma'am, we used to keep our goats in the bush. Sometimes when wild animals were mixed up with your goats, you felt honored. Domestic animals have many diseases, but when they mix with wild animals, they never get sick. If you want your chickens to be very healthy, you must bring a guinea fowl from the bush to stay with your chickens and they will not have disease.

---

*The Ndebele are a tribe in South Africa.

When the white man came, he told us that all wild animals belonged to King George, so fences were built to keep the domestic animals separate. In a country like Botswana, many wild animals are dying; they can't migrate because of the fences. In the language of the Zulu, the wilderness was called *indle.* The word means "where we get food." In our understanding, food means more than physical eating. It is also something spiritual, like spiritual food.

My son became more sick because of being rejected by his friends, and Virginia told him to go out in the wild where there are still animals. He did so and tells us now that the trees and the animals are beginning to talk to him. At first I thought my son was going crazy, but no. My son—like me—is an artist. He is beginning to paint trees so well, he says, because they speak to him. He shows me the tree that he has painted. I know that some of the trees around here are aware of us. And they miss us when we are not here—ordinary moshabele trees.

It has been scientifically proved that plants have got emotions, exactly like human beings do. Dr. Backster of America proved this in the 1960s.* A plant can be scared when threatened, just as a criminal is scared when a policeman threatens to hit him with the fist. Ma'am, living with nature is the key to the survival of human beings in the future. We were made to believe that technology could solve every problem a human can have. But no, that has not happened. Without nature, man is dead.

When man has been taught to fight against nature, he starts to fight against nature in his own house. Ma'am, you know that today there is a lot of abuse against women in South Africa. Why? Because a woman is the first representative of nature—raw nature—that a man sees within kicking distance. Why? A woman is a moon animal; she menstruates and does other "funny" things, like giving birth to children. Now, having been taught to hate nature, they [men] take their anger out on women and children, ma'am.

---

*This refers to experiments of Cleve Backster, one of the world's polygraph experts and creator of the Backster Zone Comparison Text, who administered polygraphs to plants to determine their emotional responses to various suggestive stimuli, such as threats to their well-being.

I wish to appeal to you, Your Highness: Do not play when you are teaching people to make contact with nature. Do it with all your seriousness. And when you do it, do it with all your power, because people like yourself, few as you are, are trying to save the human race. Here is healing, ma'am.

Listen, my lady. Let us look at the time of Jesus Christ, what used to happen in time: People who really wanted to experience God used to go into nature. Every city was surrounded by a thick wall, and inside the city there was no nature. The people were surrounded by the soul of this wall, and what did John the Baptist do? He no longer wanted to eat the food that was sold within the walls of Jerusalem; he no longer wanted to wear clothes. He went out of the city into the place where there was no protection from bandits; he went out into the wilderness. He nourished himself with plenty of vitamin C by eating honey out there and the fruits of the wild.

What you are doing is not new; people of all countries have done it before, seeking liberation from the confinement of people who hate nature. What did Buddha do? He left his father's kingdom; he rode out into the bush—unprotected except by the mighty forces of God. He taught people that you can leave that city and still become a human being who is one with God. If I am hungry, I go by Mister Tom's shop and buy bread. In other words, I am protected. But what if I go out, out, out, out, and there is no Mister Tom with his bread and tea? What will happen? For the first time, I will have to listen to the trees, to the birds. I will have to find food out there or die. In the end, I will find that nature is feeding me, that those things I thought were poisonous are actually my friends. For example, there is that ugly thing growing just there on the side of the road, stinkblaar. That will help me to breathe: If I burn it, my asthma will go away.

I suddenly find that the things that I found important are not so. I find that a tree is not something that I must chop down, but something I must try to protect. By protecting that tree, I am protecting myself. I am not allowed to tell of the many people I have healed of many sicknesses—including AIDS—but I wish to tell you this: Sometimes, bringing a person back into nature will heal him more than a thousand medicines will.

So we are protecting the trees against only people? There is no other enemy?

CVM: That is what we were taught: The purpose of humanity is to protect the world in which animals and people find themselves. To protect them from what? Sometimes the Earth itself starts to fight against life; it can give life and take it away. In Africa, still now, a woman can do nothing wrong. That is why when there is a crisis in an African country, the woman becomes the leader.

But you say women are mistreated . . .

CVM: Women are now mistreated in Christian countries. Saint Paul said the woman must subordinate herself to the man. That did the damage. Carrying a marriage certificate puts a barrier between a man and a woman, because she is then seen as man's property. This is not part of our culture. We had no token of marriage; it was an equal partnership between a man and a woman. With the coming of Christianity, women were put in an inferior position. A marriage certificate is like a dog license. The dog then belongs to the government.

In the animal kingdom, mostly it is the female who is the leader.

CVM: Exactly. The female gives life and takes life.

We sip rooibos tea, and Helen, the friend who came with me to see Credo, asks if this old African knowledge can be preserved.

CVM: No, ma'am. The forces against us are too strong. You must not only write books, ma'am; you must speak on television and radio. There is so much utter rubbish on television. And what is worse is that those who are doing the rubbish think that they are doing a very great thing.

If we want to do things of great seriousness, we must show people what it means to be with [what] we preach. Me, ma'am, I have tried and succeeded, but because the enemies facing me are so big, my work was destroyed again and again. Information is worth a lot of money. My name is making money for people whom I don't even know. Black people who write, paint, or sculpt die poor. I am all these things.

To come back to our theme: We can't discard the technical discoveries and the way we live at this time and age. How do you think we can combine being in touch with the oneness again, in our time, with all we have created?

CVM: Your Highness, we can give technology a human face. Who are the most powerful nations in the world today, and why?

It depends on how you see power: as inside power or outside power.

CVM: Power is the ability to command millions of people outside your country.

Then I would say the United States of America.

CVM: No, ma'am, America is not powerful. It is a weak, corrupt country, like ancient Rome. When a nation is militaristic and believes in bombs and swords, that nation becomes weak. America is weak. (With great emphasis) Her power is illusory.

May I point out, with respect, that the German people are today a very powerful nation? In fact, the Germans have gained more by peace than they could have gained by winning World War II. May I also point out, Honorable One, that Japan today is a giant among nations? Israel is a powerful nation. And why? England is powerful—so powerful that it actually tells America what to do. England today holds millions of Africans in servitude through the Commonwealth, which is one of the cruelest organizations in the world. England, Japan, Germany, and Israel are very, very powerful. The Japanese have not thrown away their traditional culture; they have carefully married the samurai spirituality of ancient Japan with modern technology.

When I look at India and at Africa, I cry. My people are total failures. And yet India, which underwent the same brutal colonization as Africa, today has launched satellites into orbit. India has atomic bombs that not even America can fool around with. Why? Because India used her great spirituality. You know, ma'am, God is a very funny thing. Sometimes out of great evil God creates great good. In India there is an apartheid system that has lasted for thousands of years. They call it the caste system. It is a most evil thing but, oddly enough, it is this that has made India the strong nation that it is. In India, everybody knows his or her place. And by being different, all people become powerful. All our fingers are different, but if you make a fist, you can knock Mike Tyson down to the ground. Out of diversity, God will make a great unity.

I have traveled to many parts of the world, and what I saw in Japan was very strange. We saw a man and a woman with a little girl approaching. They took off their shoes and went into the water of a river. We were sitting nearby. We thought they were swimming, so we did not give them any attention. But these people all died! The man had arrived five minutes late at his job with the result that production in one machine was crippled. Rather than going before the boss and being shouted at, the man went back home, collected his wife and daughter, and all three drowned themselves. No noise, no scream, no performance—just that amazingly silent walk into the darkness.

I asked myself: Why? The manager of our hotel said it happened all the time. Japanese people who arrive late at work kill themselves. We call people who kill themselves fanatics, but it's more than that: It takes a lot for a human being to walk into the darkness and die. It takes a pride so fierce that no words can describe it. You see, ma'am, the Japanese people are very productive because they are very, very, very proud. We lack that pride.

Look at the English: They are a very proud people—so much so that they sometimes are inhuman. Pride can make a man into a god and he then can do great things; but pride can also make a man into a demon.

The trouble with being old like me is that you remember too many things and forget nothing. I remember, ma'am, that in the 1930s, the Afrikaans people of South Africa were a poor and defeated people. Many of today's farms were owned by black people at that time. Then, when the Nationalist party came to power, it turned the defeated people into a nation of gods. The pride made them so strong that they were able to rule South Africa for forty years and develop some of the richest men on this planet. The men who are leading South Africa today are failing to give us pride.

Do animals have pride?

CVM: Yes, they do. A dog will die defending its own. A lioness will sacrifice her life if her mate has been shot by hunters. Put animals in a cage and they fade away and die because they have lost their pride, their freedom to run around in the bush.

Do you believe there is awareness in all life-forms?

CVM: Yes, ma'am, yes. People make the mistake to think that only human beings are aware, but they are wrong. Animals of different kinds have a deeper awareness than human beings. Why do I say so? Anyone who has looked after cattle will have noticed that if a cow is sacrificed in the village and the dung of the animal [is left] lying there out in the open, the other cows will come and bellow over the dung. They do this for some days. Our people say "to bellow over dung," which means to cry over something about which you cannot do anything. Wild animals are aware of death.

I can't help but think of Gareth Patterson's amazing story about the lion who mourned the death of her mate.

CVM: One time, I saw that when a lion was shot, the lioness came out of the tree where she had been hiding and attacked the men who had shot her mate. The lioness died, leaving a cub behind. Thus the whole family died. The lioness had been safe while hiding in the tree, but when she saw her man die, she wanted to die with him. So, what does this tell us? That this animal has feelings more intense than those of the human race. A human being would hide, but the lioness did not.

Maybe the lioness knows she has the power to attack the human, but a man is afraid of the other man who carries a weapon.

CVM: No, ma'am. Animals are aware of guns. I have seen farmers shooting birds. Sometimes when a farmer appears holding a gun, birds will fall out of trees and pretend to be dead in the grass. What are they doing? The birds are aware that a gun is a deadly weapon. [When a person carries] a stick in the hand, they don't react. Why? Guns smell of gunpowder and have a hole in front out of which death comes. Also the sound of the gun—the animals and birds associate it with seeing a fellow animal dead and they make the link.

And a gun is, of course, unnatural, while a stick is not.

CVM: Yes. We could talk on about this for hours. We are not the only creatures who feel love, who know what death is.

I tell Credo that on the nature reserve where I live, there was an eland bull attacking people and we were not safe walking around. He was probably just protecting his herd, but nevertheless, he was a real danger. We had the choice to shoot him or transfer him elsewhere for breeding purposes. I decided on the latter. He would at least be alive. We shot him with a sedative dart, and before he fell, he looked straight at me and I read in his face: "This is not meant to be." In the van, he died because he could not get up. He choked. I will never forget that sound. It was horrible and I could see, on a very personal level, how cruel we are to the animals, dictating their moves. I was devastated and knew it was all wrong. The females came and sniffed him, though they had paid no attention whatsoever when he was darted. I was amazed when the herd came up and surrounded me later that afternoon. It felt as if they were communicating that it was all right. I was profoundly touched and I could only repeat to them over and over again how deeply sorry I was.

CVM: Darting can also kill the animal because of the pain. A dart hits with great force. I have seen rhinos die from darting.

In my conversation with Rupert Sheldrake, I asked him as well if all life-forms have awareness. He answered that as a scientist, he could not say they had, but he could say there is a soul in all life. Jane Goodall, for her part, was surprised by that answer and wondered how he chose the word **soul*****, rather than*** **awareness.**

CVM: Fire is fire anywhere in the universe. So is soul. A dolphin will recognize a friend and show more affection to him. People are too philosophical. Everything has a soul.

We say that in the bush you must walk with humble feet. The bush owns you, not the other way around. I visited a mountain, the Waterberg, and if you sit at sunset, under certain atmospheric conditions you can sometimes hear voices speaking in forgotten languages. You can hear gunshots from forgotten wars and horses running. We say, "The mountains, they remember." The mountains recorded those sounds and kept them.

The animals that carry horns know their shape to precision and how to use them. I can tell you many things I have seen in the forest. We as human beings are unique in only one way: We have got hands, which give expression to our thoughts—but we are not better than the other animals. I think animals are better than we are.

I know my story will never be told exactly, but I can tell you that animals have got intelligence. Animals are changing, even as we speak, because of what we are doing to them. The animals are changing their habits in order to survive.

You are saying that all life-forms have awareness and soul?

CVM: Yes. Why does an animal show love to a wretched, dangerous, evil human being? Have you seen the love that a cat and a dog show? Even a lion will show love to a human being. Years ago, I worked underground in the mines. We think a rat is a dirty animal, but a rat can save human lives. Let me tell you: In our hostels where we used to stay, all together, we used to catch the big rats and put them in our food bags to carry them down with us and release them in the mine. Why did we bring the rats to such a terrible place? The answer is: to save our lives. How? Rocks fall down in the mine and can block the way out. About fifteen minutes before this happens, the rats know and escape. When we miners were eating and found that Mister Rat was not there to get the crumbs, then we knew it was time to run where the rats were. We could see their little red eyes waiting for us where there was air. And the rocks would fall behind us.

We should accept the intelligence of animals. We think we know so much when in fact we know nothing. We should accept that consciousness is in all life-forms, even in cockroaches. We consider them dirty, but think again: If we spray them with one sort of spray all the time, there comes a moment when they don't care any more. But bring a new spray and they run away, even before you use the spray. Somehow, they know that the new thing that is coming is dangerous.

Would you say that a stone also has consciousness?

CVM: Oh yes, it has. Granite is a stone with feelings.

And water?

CVM: I wrote to Dr. Gander,* who makes water alive. Africans used to do this to water before Gander's time.

But what was wrong with the water in those days? Is it because of today's tubes and plastics that the water loses its power? Perhaps it lacks contact with the stones and the earth and the sun and the air; it loses its natural swing and rhythm.

CVM: Water is very sensitive. As sangomas, we have rituals in which we drink water in a very special way: You drink it and then hold it in your mouth. If the water is alive, it will have a taste like ice under your tongue. Then you know you can use this water to heal people. What we also do is take the water to our nostrils and inhale. If the water gives a feeling of spinning, like blue, red, green spinning, then we know that that water can heal people. A river where the cattle or wild animals always cross: You will know that this water is alive. It is the animals that keep the water alive. But a river over which cars go all the time is a river that dies. The Botswana people gave the cars a very adequate name in the 1920s: se-ja-na-ga. *Seja* means "the one that devours." *Naga* means "the wilderness." Another word is *sefata-naga*. *Sefata* means "the one who undermines."

If a person suffers from cancer and you use living water to try and heal it, the cancer will react. For example, if you go to Drakensberg Mountain of Natal, that water will make your patient better. To heal people, we need the good water from a natural spring that is alive. You

*Johann Gander revitalizes or energizes water by transforming it with high-frequency oscillations.

will be amazed with the result: It seems to speed up the power of the medicine. Even borehole water is dying very fast now. I wonder why. Within the space of ten years, all the Bushmen in the Kalahari will be dead because of this; all the crocodiles in the rivers of South Africa will be extinct. This is a very great tragedy.

South African scientists (through organizations such as the Gondwana Alive Society)* are speaking of the Sixth Extinction, which they say is to culminate in ten years.

CVM: This extinction will come much sooner, Your Highness. Human beings [exist] in unhealthy concentrations. When people live in fear, they breed more, as do the trees and all of nature. Zulus knew this . . . We need to create a safe world, a world without anger. We can do it. We can do it! We need a more just world. We need to create true justice. It can be done.

We can only do this all together, everybody taking on his or her little bit.

CVM: In about ten years' time, cars will be driven by hydrogen . . . The Arab world will go bankrupt. South Africa holds the bridge to peace in her hands. We have platinum.† [Its] impact on the greenhouse effect will be enormous.

Credo, what do you consider to be your mission?

CVM: My mission was to make this country smile again. But I have lost two daughters to AIDS; I have lost two sons. My wife, Cecilia, was murdered in a South African hospital. Right now, we are a nation without pride, without hope. The coming of AIDS has changed the character of my people.

Still, South Africa has got the potential to become a world power.

---

*Gondwana Alive seeks to promote biodiversity and to stem the tide of the Sixth Extinction. See www.gondwanaalive.org.

† [Credo refers here to the use of platinum to facilitate the reaction in hydrogen fuel cells. —*Editor*]

I say that South Africa is going to decide the fate of all humankind, one way or the other. Our country used to be rich in agriculture, minerals, everything! We had it all. The country called Angola has become so devastated by war that people can't survive anymore. There was a time when you could eat from one side of that country to the other. In Uganda, you could eat tropical fruit all around the country—mangoes, bananas, avocados. Now, take Mozambique, Your Highness: It was a powerful country. It has now been so raped by war that there are places where you come and you listen, you smell . . . No smell of animals, no smell of even the little buck, no smell of anything. You don't hear a bird singing in Mozambique. That country will never recuperate.

Yet, Credo, there is such a powerful energy in the earth of this continent that I would not be surprised if it could change the world, even save the world, with South Africa showing the way. It is a make-or-break energy. Everything is strong here.

CVM: Yes, I fully agree. South Africa should be saved because she is the link among all the continents. She joins Australia; South, Central, and North America; India; Tibet. I went to Australia and looked for the black people of the country, the Aborigines. I found people who had deep spiritual, cultural, and linguistic connections to Africa. I found that many Aboriginal tribes speak African languages . . .

The Africans once dominated all the countries in the world. I could write a book about this. When my late wife, who was a mission school teacher, and I visited Japan in 1986 we found that many Japanese did not know English. We tried speaking different languages and succeeded in speaking to them in the Venda language, which is spoken in the Limpopo, the northern province of South Africa. I don't know of anybody else making this discovery.

Venda is one of the oldest languages in Africa. It is very similar to Congolese, to Shangaan in Mozambique. Women are the best communicators. I wish my wife were still alive. But I am alive, and I need to fight for my country. I need to speak to the world. But I am black. And black people are tied. There are white people who will not allow me to reach that platform. I have tried again and again. If people think this is a peaceful and free country, they are bloody liars. And if we do not stop AIDS, it is going to destroy many nations.

Do you feel lonely in your fight for justice and peace?

CVM: Your Highness, people are made to become terrorists when they are driven into a hole.

And if they are made very angry.

CVM: And they are made very, very angry. I try to keep people alive who have cancer and AIDS. What is the most expensive crop in South Africa? Sutherlandia. It has been commercialized for world export, although we need to use it here to treat our sick people. There are unscrupulous people who profit from prices that are far too high.

What is our relationship with nature today?

CVM: Humans have been torn away from nature. The whole planet has become sterilized. (Gesturing with his left arm toward the world outside the window) All this is dead. Animals should be jumping around. If you take a child from Soweto, ma'am, and you show him a cow, you milk the cow, and you offer the child a glass, he will say, "That is horrible." But if you bring him milk from the supermarket, he will love to drink it. We have become "denatured." We have been denatured in our spirit. When I walk past that tree, I must feel that I am walking past something that sees me, feels me, something that I need to be whole. The Christian, the Buddhist, and the Hindu religions all have become denatured. They should be either destroyed or reformed. We have been denatured regarding our own children. Why do people abuse children? Because they don't know them. They look at a child and think: What is this thing? I don't know it. And sometimes they say, "It is my child and therefore I can do anything with it." And yet in African culture, we were taught that it is not your child: It is something given to you by the gods. That is why we call a child *ngwana,* which means "a gift from above." In Swahili, they call a man *moto* and they call a baby *mototo,* which means "double a human being." We believe that babies are greater than adults; they are sacred, they are precious, and they are to be worshipped. In the Zulu language, we call a baby *umntwana,* which means a "prince," a "junior chief." We worshipped women. Some of our greatest military leaders were women. Today, what has happened? Women are maltreated everywhere.

What price are we all going to pay for that? Islam is a woman-based religion. Even the word Allah is actually the name of the Goddess. The symbol for Allah is the pregnant moon, and the star is Venus. It is the worship of one of the oldest goddesses of the world, Ishtar. And what characterizes such a religion? The readiness of its followers, the commitment. Bush has not arrested bin Laden and he has made him a greater martyr, a greater hero than Muhammad.

Where are we heading?

CVM: One cobalt bomb can destroy all life.

But so many wonderful people are here, caring, working, sending out light. There is such a strong energy of love all around us—to begin with, in nature herself.

CVM: We need to do something concrete. Why did the ancient Egyptians build the pyramids? Why did the ancient Greeks build their temples? Why did the peoples of Central and South America build those amazing structures? The human spirit must rise, because if the human spirit is kept on a certain level, without rising higher, humans become destructive. We need to keep our creative, positive spirit. Why did King Solomon build his temple? So many people can see it: If a man wants to do wrong and looks up, he sees the temple and something inside him will tell him, "Don't." Michelangelo was a crook but became one of the greatest artists of his time. It was the Renaissance! Look at his *David,* that beautiful celebration of the human body.

In every human being there is a creative spark that enables her or him to make something utterly beautiful, in whatever form.

CVM: Yes, but wait: What is wrong with our time? We are too controlled. I sit here, eighty-one years old. (Pointing to his library shelves) From the money for these books I could have created a university, because I deeply believe in learning. I believe man is created by two things: by art and by learning. But I am black, and the money I should have earned was stolen from me. I have been cheated so many times. In South Africa today, it is still difficult to create traditional African music or other art.

My last question: What is love?

CVM: Nobody knows. It is a form of self-sacrifice. When a woman sees that a certain man is a monster, a rogue, and should be thrown in jail and kept down in that hole, but she believes that she can make him better, that is love. Love is the arm of God. There are things in this world that make us into gods.

You can describe the screaming of airplanes, the clattering of tanks. You can describe the cries of soldiers, the drumming of machine guns. But love and peace are indescribable. In old Africa, you were not allowed to use the word *love;* it was too holy to pronounce. If we could say the word *love,* the whole universe would dance.

Works by Credo Vasamazulu Mutwa

Indaba, My Children: The Journey to Asazi. New York: Grove Press, 1999.

Isilwane = The Animal: Tales and Fables of Africa. Cape Town: Struik, 1996.

My People, My Africa. New York: John Day, 1969.

Song of the Stars: The Lore of a Zulu Shaman. Barrytown, N.Y.: Barrytown, 2000.

Zulu Shaman: Dreams, Prophecies, and Mysteries. Rochester, Vt.: Inner Traditions, 2003.

DENISE LINN

I believe that it's damaging for the soul if we separate ourselves from nature.

One glorious morning in Washington, D.C., we have time to visit an exhibition on the Egyptian tombs at the National Gallery before catching a plane to our next interview. It is a special treat for me because I feel a deep kinship with the goddess Isis. As the legend has it, her husband, Osiris, has been murdered and parts of his body have been scattered throughout the country. Isis gathers the pieces, reassembles his body, and rekindles it to life, after which she and Osiris conceive a son, Horus.

Isis, to me, represents life force: the possibility of a new start after a conflict, pain, or illness; new life where all seems lost. Again and again, I have witnessed how true this is, and it has helped me to focus ever more on "the possible," which I find enhances a deeply loving quality of life. By focusing on life, you focus on love, which surpasses death. This is what Isis has taught me.

Rushing to the airport a few hours later, we find that our flight is canceled. "Technical problem, sorry," we are told. But nothing can spoil our excellent mood. We are having such a good time, it seems to be contagious—with a smile the ATA employee transfers our tickets to a later flight to Los Angeles! Jessica and I curl up on the very last seats at the back of the plane. The delay gives us more time to rethink the book, the sequence and content of the conversations. We catch one more small plane in L.A. and our energy is still light and happy when finally, at midnight, we arrive in San Luis Obispo.

Tall and striking in blue jeans and sweater, Denise Linn stands there to welcome us. Long, raven hair frames her strong, lovely features; her piercing green eyes constantly change from reflective to sweet to laughing, revealing her inner depth.

I had read all ten of her books and found her most recent work, *Secrets and Mysteries,* at the Johannesburg airport. She uses her writing to come to terms with issues in her own life or to explore a theme dear to her heart.

We drive to her home through the moonlit night. Once we arrive, sleep overcomes me and I collapse on the bed, exhausted from our travels but also from the five-hour time difference. Jessica continues on up the hill, where a little goat house converted into a Buddhist-style guesthouse awaits her.

The following morning brings a brisk and bright quality to the day and, wandering around the grounds, we discover a pretty yet simple and energetically well-balanced house. So this is what Denise is about. It is an oasis of peace and welcome, stretched out among rolling hills and valleys, encircled by impressive four-hundred-year-old oak trees, a beautiful persimmon tree, some newly planted fern trees, and a pretty garden that calls for constant watering in the hot California climate. The old oaks are honored by the name of the place, Sacred Oak Ranch. Pepino—the sweet stray dog Denise has taken in—greets me, leaping up and licking my legs.

On an enclosed patio filled with round tables and reed chairs, our hostess has set out a breakfast of delicious fruits of all kinds, buns, and honey. While talking to Denise's husband, David, and Rafaela, the computer whiz who is helping out, we are surprised by the sound of something hitting against a windowpane. A goldfinch, small and delicate, drops to the floor. Denise takes the tiny yellow and gray bird ever so carefully in her hands, feeling its condition. She talks to it inwardly, sending it healing power, yet knowing that the bird will be the one to decide if it will live or die. Then she leaves to place it in the garden on the grassy earth.

What does this incident tell me at the start of the day when we are to interview Denise? That Denise's life is connected to the choice between life and death as much as to a balance between the two? This strong and vulnerable woman radiates both inner and outer harmony, a peace she has obtained through an extremely difficult life. She teaches that we can all heal our wounds and find inner peace, no matter what hardships we have endured. I think of Isis again and I understand then that our visit to the exhibition in Washington was the real beginning of our visit to Denise Linn.

For the interview, we sit around a comfortable coffee table on wooden benches with big white cushions and pillows. The grapes that grow over the pergola provide a welcome shade; the wind plays softly with chimes hanging from the pergola, producing sweet and peaceful sounds. It is a beautiful and safe place to speak from the heart.

What is nature?

Denise Linn: That's like asking "What is God?" . . . To me, nature is the life force within all things, manifested in trees, stones, the earth, the sea, in us. The planet is flowing in a way that is natural and balanced, according to its own nature, its own pattern. I think everything has a rhythm and its own unique pattern. So, to me, nature is living in accordance with those patterns.

It isn't easy to answer your question, because it is like trying to put words to something that is more experience than anything else. For me, nature is how I feel when I am in a place that nurtures the soul. It is an experience that reminds me of who I am. And when I am in those natural places and stop to quiet my mind, it is a remembrance almost of where I come from, of who I am—of where I am going and what my role is in all of this. (Pause) It's interesting, because I feel that I am trying to answer this question for myself, listening to what I hear inside . . . What is nature to you?

Similar to what you say . . . your words are very recognizable. It also has to do with space, the space that every being needs. It is something from the inside out. It's not a static thing, more a process.

DL: Yes, it's a process, an experience, a feeling, an awareness. And even deeper than that, it is *to be*. That's the essence. My experience with human beings is that we like to separate. There's value in that, because it allows us to be distinct. But the difficulty with that concept is that we lose the wholeness of all things, the connection between it all.

How does nature play a role in your life, Denise?

DL: When I'm truly in nature, and that means that I'm not thinking about something in the future or in the past, it's a reminder of what is true, of who I am. It's a reminder of my place in the destiny of all things. It takes me home and allows me to feel at home not only with myself and with others, but also in the universe. The challenge, I find, is to actually be in nature when you are there.

I've been teaching for about thirty-three years. I've taught many different things over the years. But when I look at what is beneath everything I've taught, nature has been the central issue. But my work eventually evolved into feng shui. And you might think that this just deals with inside the home. But, in fact, my focus was on how to create a home that has the same feeling that you have in nature. How can you activate the same energy, the same feeling, the same life force in your home that you experience when you are in nature? So, nature has been a theme that has run and is still running and will continue to run through everything I'm doing, because I believe that it's damaging for the soul if we separate ourselves from nature. Much of what I've done is to create a context within which is a natural reconnection to what supports us and nurtures us and heals the soul.

All my questions are answered. This is going to be a very short interview! What does nature teach you?

DL: When I'm in nature, there's a raw honesty. It teaches me so many different things. It teaches me to be honest, for example—not only to other people, although that is something very valuable, but it teaches me, as well, to be honest to myself. That means to really listen to the secret messages in my soul. When I'm angry, for instance, I have to accept the fact that I'm angry and not suppress or deny that feeling. Only then can I be completely honest to myself.

It also teaches me to recognize that I'm part of everything else. I think that in our modern world, there's a sense of isolation and separateness—not only from nature, but also from each other, from a creative source, from God. Being in nature reconnects me to God, reconnects me to other people, to the Earth, to the universe. So it gently removes me from the confining and defining world and immerses me in stability and interconnection.

And having a house that is in harmony with nature means that it is easier for people to stay in harmony with that source?

DL: I know that not everybody is fortunate to be able to live in nature, but it definitely helps. I do feel, though, that there are ways we can implement that natural feeling into spaces so that they become more humane.

One very simple thing is the lighting. You know that most offices have incandescent or fluorescent lighting. Plants can't live under just incandescent light, because the light spectrum is mostly in the yellow range, whereas out in nature we have the full spectrum with all the colors of the rainbow. And studies have shown that spending time under fluorescent light contributes to calcium loss from bones and teeth. Yet we put ourselves in environments in which not even plants can live! A simple solution is to put in wide-spectrum lighting.

For me personally, I am blessed to be in a place where I can reconnect not just with nature, but also to the native cultures. Native people talk about deep connection to the energy forces of the earth. They believe that those who really listen to the earth can receive that connection, that energy.

I believe that whatever you love in life responds. Even driving your car and giving it some extra attention might actually help it to work better. So, by talking to the earth, by acknowledging and honoring and loving it, it is my belief that the spirit of the land will nurture and support us, because there is that connection and communication.

I tell Denise about Patricia Mische's concern that when she is teaching, she doesn't have time to go out into nature. *(See pages 270–71.)* *She walks through a park, to and from the place where she teaches, and that's it. But while listening to her, I understood how she is bringing Earth into her peace studies just by walking to and from work. Unconsciously, she makes the connection. It was a completely new thought for Patricia.*

Denise responds with a sigh, apparently happy to be reminded of how the energies in nature touch us without our realizing it.

In an earlier interview, Jane Goodall told us she doesn't need to go out into nature anymore because it is so much within her, while Patricia said she feels the need but doesn't have the time. That is when I thought: But you are connected, being with your students all the time.

DL: Oh, that's nice to hear, because all those years that I was teaching, I was talking a lot about nature and yet I spent most of my time in airplanes and hotels! And often the hotel rooms I was teaching in didn't even have windows . . . so, often I would think: What am I doing? I'm teaching about nature and yet I'm doing it in such sterile environments. That's why it's nice to have that perspective that you just mentioned.

It reminds me of this amazing study done in hospitals. They wanted to know what can influence recovery after a particular surgery. So they divided a group of patients into four smaller groups, knowing about how long the recovery would normally take.

The first group underwent regular treatment. The second group was recovering in a room with a beautiful view out to nature. The third group they put in a hospital room with a picture of nature. And the fourth group recovered in a room with modern art. The results were very interesting: The first group recovered as fast as expected. The group with the view out to nature healed dramatically faster. But even more remarkable was that the patients in the third group—with a picture of nature—healed as fast as the patients with the actual view! And the modern art people actually took longer to recuperate. Unfortunately, I forgot the reason for that.

So maybe Patricia Mische, while talking about and teaching nature, is getting images of nature in her head, causing a connection—

just like the patients with the pictures of nature in their hospital room.

I was in Sweden once, where there was a big geometrical picture hanging from the wall. I caught a cold while I was there, and I would sit in front of that big picture and it would make me feel better. I told the people at the seminar how good it felt to just sit down there. I said that I didn't know what it was, but that I felt that the symptoms of my cold were going away. They said that it was interesting, because it was a photograph of the crystalline structure of vitamin C!

The funny thing was that I really didn't know what it was, but just felt better sitting in front of it. I responded really strongly.

Now that Denise has brought up healing, it is easy for me to raise the subject of her healing work.

Have you always been a healer? How did you discover it?

DL: I think it became more activated when I had a near-death experience at seventeen. When I came back into my body, I realized something had happened on the "other side." Even while still in the hospital, I began to hear music that other people weren't hearing.

I was taking my wheelchair out on the hospital lawn once and heard this tone. (Actually reproducing the sound with her voice) It was this beautiful sound. I tried to detect where it was coming from, and looking at the grass, I realized that the sound was animating the grass. Then I noticed a tree on the other side of my wheelchair [making] a different sound—(trying to dig the tones from her memory) and with an incredible light around it. It was almost a vibrational sound of the nutrients coming into the tree from the earth. I could even hear the sound of the sunlight being absorbed into the leaves. I could see that energy exchange. And as the wind began to blow, I heard the leaves making a beautiful crystalline sound.

I know ancient and native people have always been sensitive to that energy. And I believe that my near-death experience somehow opened that door inside of me. All of a sudden, I could begin to sense and feel energy in life-forms. I began to realize that we live in an ocean of energy, and in every moment we are taking energy in and letting energy out. I believe that this experience began to validate my future work as a healer. I eventually became a healer for others as a result of healing my own body after the incident that resulted in my near-death experience. I had a lot of injuries: I lost my spleen and a kidney and adrenal gland. I also have a plastic tube replacing the aorta [and] my small intestines and my stomach were damaged.

What are you saying? Did you have an accident?

DL: Well, I was only seventeen years old and had a little motorbike and I was out for a short ride. Evidently, I was followed by a man who was actually a sniper, who killed a number of people. He took his car and rammed into my motorbike, and when I was trying to get up, he came back and shot me. There were some other things that happened before he left, but then eventually I was left close to death on the side of the road. Someone found me and took me to hospital.

That is how I lost all those organs. The doctor said, "There's no way that she will live." I had substantial damage, a big hole in my spine that looked like a cartoon hole . . . and I was told I would never have children.

But I believe that something happened, in those moments when I was thought to be dead, that changed my life and activated a healing ability in me. You see, I think that we all have an ember inside of us that's a reminder of who we are. But sometimes that ember is very small and covered with the ashes of forgetfulness. However, I believe that during those moments when the doctors thought that I'd died, it was almost as if the warm winds of heaven blew gently on that ember and it became a little flame that ignited a healing force inside of me. As a result of that, my body recuperated much more quickly. Everybody says it was a miracle. I don't think it was a miracle; I think we all have that life force that makes us able to heal.

Do you remember what happened during the time that the doctors thought you were dead?

DL: Yes, I do. I'd just been taken to the hospital and I was in a tremendous amount of pain. Suddenly, all the pain left and I floated out of my body into a place that was filled with golden light. It was so beautiful and light and serene and filled with love, but also so familiar! As soon as I got there, I knew I'd been there before—and it felt more real than anything here did. It was the most real experience I'd ever had. So, this earthly life all of a sudden felt like a dream. I couldn't conceive of the past or the future. Everything seemed to be timeless. I remember trying to think of the past, but I physically could not do that. I had an awareness that I had been on Earth in a place with a past, present, and future. But there, I couldn't conceive of this.

Back on Earth, I can't seem to experience this timelessness anymore, because I am hooked in this time-space continuum. But there, I couldn't imagine the past or the future. That was interesting. When I was on the other side, I had the feeling that everybody was there: everyone that's ever been and everybody that's ever going to be. We were all there together, individual yet all one. I remember the experience, but I can't experience it here.

The most amazing thing about being there was that the infinite love and the light and music weren't separate; they were all merged together, they were all the same. And this life here just faded . . . Well, it was really hard when I came back into my body. I couldn't talk about it without crying hysterically, and I felt this deep, deep homesickness. In fact, I got better at holding back the tears, because I have talked about it over the years. Now, I remember having talked about it instead of remembering the experience itself, and that's been useful, because when I actually remember the experience, the yearning is so deep that I have a hard time. There is this sense of loneliness, isolation . . . separation. (Fighting tears, as if this is one of those times when she is actually reliving the experience) When I came back in my body, I really wanted to go back there. However, I knew that suicide was not an option to go back to that place that some people call heaven. I also had the feeling that heaven wasn't up in the sky. (Pointing to the sky) People think that if you die, you go up there. But it seemed to me that we can touch heaven even while we live on Earth.

I'd read about people who did Zen Buddhism. Their experiences seemed in some ways similar to what I'd experienced, so I decided to enter into a Zen Buddhist monastery, where I lived for several years.

Sometimes I sat in meditation for sixteen hours a day. It was tough and at times very painful. But it was a special time. I didn't reach heaven, but in silent meditation there were times when I felt very close.

Where was this?

DL: The monastery I moved into was in Hawaii. It was connected to a Zen temple in Kamakura, Japan. The monks and Zen master used to travel between their Hawaiian monastery and the one in Japan.

When you can sit, facing the wall, and watch the shadow of a leaf falling across the wall and find an explicit joy in that, it is a remarkable experience. It was a simple place, nothing elaborate or fancy. I liked that.

After several years, I moved out of the Zen monastery and eventually entered the next phase of my life, which was to be a healer. My substantial injuries from being shot had healed quickly, but my injuries also ignited an interest in healing. My first teacher was Morna Simiona, a Hawaiian *kahuna,* which is something like a medicine woman. I found her in a most unusual way: I was looking in the phone book under *massage,* and just picked a name. The masseuse I picked worked in the tourist town of Waikiki, in the basement of a hotel. I went to the hotel and waited for my massage. When it was my turn, I looked at this older Hawaiian woman who was to be my masseuse and started crying. I couldn't stop crying. I was embarrassed, so I apologized.

It turned out that this woman was Morna Simiona. I had had such a profound experience with her that day that I wanted to train with her. She finally accepted me as a student because of my Native American heritage. She thought that was important. She talked to me about the secret messages in the trees, the clouds, and the stones. She talked about the elves and the fairies, and she would collect big bowls of fruit to give to the *menehunes,* who were the elves of the forest. I was standing at the edge of the forest, acting as her "lifeguard." She was such an honest, humble woman, yet she talked about fairies. I would find the schism with my Western mind. On the one hand, I couldn't see any elves, yet on the other hand, I believed her when she talked about them.

Many years later, I had a call from a Native American from New Mexico. He had seen an article about me that he'd kept for a year, during which time he and the tribal elders had sent prayers my way from the *kiva*. Eventually, I met him and he became my teacher too. He was a remarkable man who would talk about turning into a fox. My Western upbringing made me find this hard to believe, yet he, as well, was such an honest, humble, soft-spoken man.

Again, I would feel the schism between my Western upbringing and the native traditions. Yet because of my near-death experience, I knew that there was truth in what he said, and I believed that he would become a fox.

Because you had experienced the oneness of all life . . .

DL: Yes, I knew that. And yet, sometimes there would be another part in me that believed that he just thought that he became a fox. I would hold both views equally. That was sort of strange.

What part of you is Native American?

DL: On my mother's side.

And have you been brought up in her tradition?

DL: No, I wasn't raised in a traditional way. When my mom was growing up, to be Indian meant that you were scorned. Now, being Indian is in vogue, but then, you were a second-class citizen. Indians were discriminated against and looked down upon. Because my mother was fiercely proud of her herritage, she was angry and she was very interested in getting political freedom and political rights for Native Americans. But my mother was not interested in the spirituality [of the tradition], so I wasn't raised in that spiritual tradition. I was raised more with her anger. But then, you can understand that anger, when you understand the history of the Cherokee people.

When white men came to America, the Cherokee were forced off their land, which was then on the East Coast, and they were forced to walk across many states and many miles to the dry, arid land of Oklahoma. This was extremely difficult for my ancestors. Many people died on that walk, which is called the Cherokee Trail of Tears. Before they were forced off their land, they were farmers, unlike the Plains

Indians, such as the Apache, who were nomads. The Cherokee had lush, fertile land on the East Coast in Georgia and South Carolina. They were also miners; they had gold. They also had a strong sense of democracy. They began to intermarry with the whites, because they felt [themselves to be] equals.

They had their own newspaper, their own language—they were the only Native Americans who had their own written language. They even had representation in Congress. Most people don't know that, and call them savages.

Then one morning, they were told to leave. They were forced to walk from the East Coast to a very dusty land. There are reports of some soldiers who walked with them who, by the end of the journey, wrote of their respect for the courage and dignity of the Cherokee. Many people died on the Trail of Tears. They were allowed to stop only every three days to bury the dead. In the meantime, they were told to put their dead on carts and pull them. But the Cherokee would not allow their dead to be thrown on those carts. For three days a husband would carry his dead wife, without complaining, or a son would carry his dead sister. They just walked with this immense dignity. To read the soldiers' journals about the courage of these people is amazing.

I believe that the native spirit is something that's in your soul, not in your blood. So even though I wasn't raised on a reservation, I have it inside of me.

When did you know about your healing power?

DL: I think it might have started even when I was younger. Both of my parents were interested in science. They were passionate about it and didn't believe in the spiritual. But my grandmother on my father's side was an astrologer and numerologist and had studied mystical traditions. My father forbade her to talk about any of that with me!

Why?

DL: Because it was not scientific and he thought it was too superstitious.

Or did he feel that you had it in you?

DL: I don't think he recognized that was a possibility because he was so scientifically focused. But when my mother had to go to a mental hospital for a couple of years, I lived with my grandparents, and my grandmother had those expectations of me. She would say: "Denise, put your hand here, because I feel pain. I know that you can make it go away." Or she told me, "Denise, I can't find my checkbook. Where is it?" She expected me to develop my intuition—and I believe that whatever is expected of a child is decisive for the direction [he or she] is going to take.

Was it through your near-death experience that you realized the importance of the inner ear?

DL: I think it helped to bring things in my life into focus. Even though having a near-death experience sounds amazing, it didn't mean that everything was perfect afterward. In fact, my life was even more difficult afterward. But the experiences that I had helped me through those difficult years.

My mother was schizophrenic and she was violent, which is difficult to live with. And then, when she was in the mental hospital, my father was sexually abusive. But in those days, you didn't want anyone to know about these things. At all costs, you wanted to keep it hidden.

Those costs must have been high for Denise. But it seemed that after she had been shot, all that denial came up. She went through some very rough years, with suicide attempts and experiencing the darker sides of the soul. We might assume that because she had experienced the spiritual dimension of life during her near-death experience, this would have made living relatively easy. Instead, it removed the lid from a great well of suppressed childhood trauma.

Denise talks about how she moved on after these experiences.

DL: My journey to wholeness was a long one because of the emotional, physical, and sexual abuse. I can see now that all of that was part of my journey, and it allowed me to become who I am today. Ultimately, it was good. I'm still on the journey.

My mother, during her fits of mental illness, would always tell me, from my early years on, that I was worthless. She would say, "You can fool other people, but I know who you really are. Other people can like you, but you can't fool me!" As a child, I of course believed what she said.

However, I bless my grandmother, because she used to say, when she saw me struggling, "Denise, the roots of the trees go deepest when the wind is strongest." She said: "You are having some strong winds, but your roots will go deep." God bless her for that, because it was hard for me. There were months and years that I just wanted to be dead. It was hard because of my feeling of self-disgust and shame. In later years, when I met other people with the same experiences, it allowed me to show compassion and I understand things that I could not have understood if I hadn't experienced them. I don't think you need to suffer to grow, and I don't think you need to hate. But in my case, my deep emotional wounds did facilitate compassion.

Who helped you through this?

DL: My solace as a child was always nature. During the years that I was being abused, I would often sleep in the bathtub at night, with the doors locked so I wouldn't be abused in the middle of the night. But sometimes, when the violence and the arguments between my parents were intense, I would go to the woods and spend time there. I got to know every part of the woods. So my solace was always in nature, just being there. I actually think that even before my near-death experience, nature was healing for me. I remember that when the violence became so difficult, I went into the woods and found this hole in the earth, just big enough for my body. I would go into that hole, and I still remember the wonderful smell of the earth! I would feel all the pain floating out of my body.

Some people think their solace comes through other people, but my solace comes through nature.

It was not that you wanted to die there, but that you felt nature was your friend?

DL: As a child, I just felt much better there. I sometimes ate some of the dirt. I know some psychologists believe that children who are abused will sometimes eat dirt. But to me at the time, eating dirt deepened that sense of connection. Just holding it in my hands and curling up in it, it absorbed my pain.

You have this strong communication with the earth. Through your story, people will understand that your way has been the right way for you and that nature in all different forms can help you, giving you messages and energy.

DL: I remember as a child my grandparents used to live on a ranch in Oklahoma. One morning, I got up really early. The sun was coming up and I could feel that the earth was waking up. I remember having an incredible experience that morning. I could feel the consciousness in every aspect of nature. It was as if I could hear the "voice" of all parts of nature. Everything was alive. It was such a moving experience for me. I remember that day it was difficult for me to talk. I didn't want to go back to the ranch. It was difficult to even communicate. Now I know I was in an alternative state, but then it was a new feeling.

Do you feel that you are doing this work because you are still healing yourself?

DL: I think that the only person you are truly healing in the end is yourself. And even when I'm doing healing work on others, I feel that I'm healing the Self in other human beings.

I'm convinced that you teach what you need to learn, and what I'm teaching is: Who you are is enough; you are safe and valuable. That is what that little wounded child—that little me—needs to hear. I think it's fabulous that other people can benefit from my own healing process. I used to teach really large groups. Now I'm teaching less and less, and finally I'm learning what I have been teaching all those years. The strange thing is that I sometimes suddenly feel that I don't know how to teach anymore. Does that make sense?

Oh yes, that makes complete sense to me! . . . Do animals have a special place in your life and work?

DL: One aspect of my relationship with animals is that I see them as messengers. I find that there are always secret messages around us, but we are often so cluttered with mental activity that we aren't aware of those messages. But these messages from the universe are always there: in the plants, in the sky, in the trees, in the clouds, and coming from animals. The universe is always whispering and talking to us. One particular way that messages keep coming to me is through animals, especially wild ones.

Do you think they are telling us something about this world that we should hear and understand?

DL: (After some thought) There are not only individual messages from animals, but also collective messages. Each animal grouping has, just as human beings have, a collective consciousness. And I feel there is a deep communication, almost a cry for balance, for peace.

We have had foot-and-mouth disease, mad cow disease, and avian flu in Europe, Africa, and Asia. That should tell us something, shouldn't it—those massive deaths—as well as what is happening to the whales and the dolphins beaching themselves . . . ?

DL: I read that the sonar sound tests executed by the Navy are the equivalent of us humans putting our ear up close to a jetliner coming in. That is how intense that sound is for the whales.

What did the little bird tell you this morning?

DL: Well, I heard a lot of things. One is the preciousness of life. I was just aware of how precious every moment is. In my book *Secrets and Mysteries* I tell about the rough period I went through when I was diagnosed with cancer. It was tough. I asked myself the question "Am I ready to die?" And I thought: Yes, I am. Of course my husband will miss me, but he will be fine. And then the very next day, I asked myself, "Am I willing to live?" That was quite another question, and that was the difficult part!

I faced that question and it made me live more in the now, cherishing every moment, honoring the preciousness of the present. That awakening was part of the seven weeks of solitude I had on this ranch, because all of this happened before I came here.

And then the strangest thing happened. When I went back to the hospital, they couldn't find any trace of the cancer anymore. I felt double blessed. First, I was blessed with the diagnosis: It made me aware of the fact that I did not live fully.

On the night we arrived in San Luis Obispo, Denise told us how she and her husband, David, decided to buy the ranch and move their lives from the humidity and rain of Seattle to the sunshine on the central coast of California. She traveled ahead to their new home while David remained behind to attend to some details. Denise was alone for seven weeks with only one chair and a mattress in her new home. She was literally on her own; she knew no one in the area and had not even a telephone or Internet connection at the time. Instead, she spent those weeks walking on their new land, listening to the old, wise trees and all the new sounds, detecting more than thirty bird species, and feeling completely connected to it all.

She solemnly promised to spend at least one month a year in that way, so that she would be able to hear the silence and feel as one with nature.

A few days after she has told us this, we realize how revealing it is that we will stay at Denise's ranch for several days. We traveled considerable distances for the interviews we conducted for this book, but nowhere did we stay for such an extended period of time as we did on the Sacred Oak Ranch. Denise showed us the peace and quiet that was sometimes lacking during our world tour. Her house and land offer that peacefulness, and so does she.

Denise returns to the story, telling how she finally decided to surrender and then received the phone call informing her that the cancer could no longer be found in her.

DL: Being able to surrender is a precious thing in life. We are so caught up in our "to-do" lists. I was so busy trying to please people, trying to meet other people's expectations, making everyone happy. In my being there for everyone else and helping and spoiling everyone else, I completely forgot about myself!

Back to the little goldfinch this morning. I was holding him and was thinking how precious that moment was. It brought me back to that diagnosis, and how wonderfully, in a way, it can work if you have a death sentence like that. I mean, if you have a death sentence, you know that this may be the last rainbow that you will ever see, this may be the last meaningful conversation you will have, (pointing to a bunch of fresh grapes on the table) this may be the last sweet grape you will taste. All of that takes on a specialness—like holding the bird this morning and later checking on him again. At first he was unconscious, his head was kind of heavy, and he was between life and death. I felt more than anything the preciousness, even if he was going to die.

Seeing you with the bird made me realize how much of life is lived between life and death, and how you have to make decisions all the time.

DL: Yes, and you can choose which way to go.

We both take a long, silent break, each of us in our own world, yet connected by what we are sharing today. Then we talk about our fathers. Denise brings up how her father died.

DL: My father had cancer, and he didn't believe in life after death. He was just going to die and that was it. He said, when I asked him what he thought was going to happen, "Nothing! That's it. The end." At one point, though, he said, "Well, if there is a hereafter, I'll put a big *M* in the sky to show that the Mormons were right." He had some friends who were Mormon.

A few weeks after he died, I was driving down the road. There was a totally clear sky, and all of a sudden I saw a big *M* cloud. There were no other clouds except for the *M*—not just an *m,* but a perfect, big *M!* The woman sitting next to me said, "Look, there's an *M* in the sky," and the other people in the car shouted, "Stop! Look! There is an *M* in the sky . . ."

(We all laugh) It could have been a coincidence, but I don't see it like that.

How did you say good-bye to your father after what he had done to you?

DL: I sat with his body just before it was cremated. It was a powerful moment for me. I held his hand—it felt so cold—and I talked to him from my heart in a way that I had never been able to do when he was alive. I told him that I forgave him, though I think I didn't believe him when he said he was sorry. In that moment, I felt a deep love for him, because all of a sudden I realized what shame and pain he must have been living in for all that he had done wrong. Even if he didn't consciously acknowledge it, on some soul level he would.

Denise, I want to talk about the world situation: Where are we? At the Johannesburg Earth Conference in 2002, a lot was said and promised, but where has that left us? Terrorism; the always-useless wars . . . Rigoberta Menchú says, rightly, that people seem to have forgotten that two hundred thousand people died in Guatemala during the dictatorship there. We have worldwide problems to face—poverty, HIV, and AIDS. None of us lives on an island; we are all world citizens. We cannot deny anymore that everything is related. How do you look at all this?

DL: I see the world in polarities: There are light and dark, day and night, male and female. And it seems to me that the greater the darkness, the greater the light. It is almost like a natural law. In many aspects, we have seemingly entered into a time of great darkness in terms of mass destruction and mass pollution. We are massively disbalancing the Earth.

Yet at the same time, it seems that there is an awakening of consciousness, exponentially—people yearning to find out more about who they are, yearning to embrace spiritual tradition, and people in the Western world moving to religion as well as spirituality. The greater the darkness, the bigger the interest in spiritual ideas. And it also awakens the interest in helping the Earth. Even people who normally aren't that interested in things like that are starting to get interested in helping the Earth, understanding the relatedness.

About the attacks on September 11, I have to tell you about something remarkable that happened to me: I dreamed about the attack some fifteen minutes before it happened. I had a nightmare about a huge, huge tree. It was so tall. There was a woman screaming, and there was a bearded man at the bottom of that tree. At the end of my dream, the whole tree collapsed in on itself in a way similar to how the towers collapsed. There was blood everywhere. I was upset because of this dream that I got up—didn't even get out of my pajamas—and went up the hill. There I stood in the tall grass, thinking, "I must have some really deep problems to have such a frightening, horrific nightmare." I had never had that before, and haven't had one ever since. I took time to catch my breath and I watched the sun rise, because it was not even six in the morning here on the West Coast. I came down the hill, still horrified, thinking, "What was that about?" And then, when I saw what had happened in New York, I realized it was exactly what I had seen in my dream, only in my dream it was a tree instead of those towers. I'm sure I am not alone; I'm sure a lot of people had premonitions or dreams about the September 11 event.

What I found out after 9/11 is that many people began to pray, to meditate. And surveys show that a lot of people are spending more time at home with their families. There is this new sense of unity, together with this massive new interest in spirituality.

It is easy to get upset by the horrible events. There are many horrible things in the world now. For instance, our water here on our land is pure and sweet, but I'm concerned about all the new agriculture that's occurring in our area. And what about the air, the earth, animals, trees? Technology has damaged the ecology worldwide.

It is so easy to get caught up in the negativity of it all. I always think it is better to see the joy of things, although it is difficult sometimes. But I also think that, because of the numerous news reports about pollution, many more people have an awareness of the environment. If it is *your* garden, *your* food, and *your* child who can become sick, suddenly the awareness emerges that what happens "over there" will have an impact on your life, too.

If there is so much poverty, such a lack of water, diseases spreading, is talking about nature a luxury?

DL: Oh no, I don't think so! If you know how to create harmony with nature, the peace that ensues can help you have a stronger immune system. And the more vital nature is, the easier it is for crops to be productive.

There was a man named Dan Carlson who was quite shocked when, during his time in the Korean War in the 1950s, he saw a mother crippling her son so the child could go and beg for money. He was so upset about what he saw that day that he made the decision, when he returned to the United States, to become a scientist to see if his research could help increase crop production so that no mothers anywhere in the world would be faced with having no food for their children.

In his research at the University of Minnesota, he had learned about particular sounds that could open the small holes on the back side of leaves, called the stomata. When the stomata are open, leaves can absorb more fertilizer and this increases crop production. A fellow researcher noticed that the sounds used were similar to bird songs, so Dan began using these in his research and found that bird songs could dramatically open the stomata.

Dan Carlson's methods are being used today in farms around the world with remarkable results. In one orange grove in Florida, trees were sprayed with liquid nutrients at the same time that birdlike sonic tones were played through the grove. The oranges from those trees were not only sweeter, but also had one hundred percent more vitamin C than those from the surrounding orange trees. Because the songs of birds are beneficial to plant life, it is probably true that these sounds are also beneficial to human beings. So I consider understanding nature is not only a luxury, but a necessity as well.

And then there is the story about the amazing woman whose father or grandfather had the biggest oil field in Texas. She owns about seven hundred acres and decided to bring the natural balance back to that place. She is running cattle there now, although it is probably millions and millions of dollars' worth of land. But she wants the cattle to be there and is very nature-centered.

I tell Denise the same story I shared with Arne Naess: I was once sleeping between two beautiful rocks in the Swiss mountains at full moon, and had to turn away because the conversation between those two rocks and between the rocks and the moon was so intensely intimate—almost like lovemaking. Naess said, "That is how everybody should experience it—but as a philosopher I say, 'No, there is no life in stones.' "

He had a dilemma there, a battle between the provable and the knowable, or those things he felt but was not allowed to feel, scientifically. What do you think about the awareness in stones?

DL: For me, there is as much life, light, and consciousness in stones as in a tree, animals, or humans. Even in the year and a half that I have been in this place now, stones seem to keep coming up to the surface of the land. I can almost see them emerging, and it feels like a kind of rebirthing. I have had times where I got really still and just sat and could almost hear the songs of the stones. Sometimes I almost feel as if I can see the stone spirits—fairies—but maybe it's a trick of my eyes.

Is there awareness in every life-form—or is it a soul or, rather, a consciousness?

DL: It feels like a consciousness. When I mentally try to figure out exactly what it is, I get confused. When I become still in a wild place in nature and listen, I can talk to the spirit of nature, and the information I gather is valuable in my life. That's enough.

My Cherokee ancestors believed that all life was connected. If you think about it, even our bodies, which we feel a personal ownership of, are a part of all things. For example, the air in our bodies has been everywhere. There are molecules that we are breathing that Gandhi breathed before! The water in our bay has been in the sea, in the Amazon, in the snow of the Himalayas. It has been in the Mississippi River, on the bottom of the ocean. It has been in all of those places. On a purely physical level, our bodies are connected to the greater flow of nature.

There is incredible research done in Japan on the idea that water holds memory.

That is the research of Masaru Emoto, whom we've also interviewed for this book!

DL: Oh! His research is amazing. It shows us how our thoughts influence the matter around us.

As humans, we usually think of a stone being separate from us, and yet the physicality of our bodies has been a part of the earth. It has been an apple, has been a grape, a cow. And it is all from the same earth, as is the stone. Perhaps our ability to communicate with the stone comes from the fact that we have that energy inside of us. Our bodies come from the earth and will go back into it.

And awareness?

DL: The awareness of stones is different from the one you'll find in a plant. And even the awareness in a particular kind of stone varies: A river stone has a very different consciousness from a stone that you find in the mountains. The bottom line is more a question of results.

I believe that connecting with the consciousness of nature can create positive results in one's life. You may ask how listening to a stone makes a difference in life. I found that people who are able to open that door are more at peace with themselves and their surroundings when they take time to listen to nature.

It also increases the richness of life, so there are results. Sometimes, people are afraid to do this, and I tell them, "You don't have to do it, but just try it. Just try imagining you are talking to that stone—what would it tell you? Just use your imagination." And then they tell me afterward what the stone told them. That's a result. I believe this is how native people understood which plants to use for healing—it's because they talked to the plants. The result was that they had access to healing herbs. Whether you believe it or not, it will produce a positive result.

It enhances energy. This is not a placebo effect. And we shouldn't forget the fact that giving the plant our attention makes its healing energy stronger.

DL: And many times now, scientists who have done research on those native herbs detect their specific chemical reactions. But I'm convinced that the way native people found those herbs wasn't through trial and error; it was through talking to them.

Do you consider yourself a pioneer?

DL: No, I consider myself a good student of life. Much of what I have learned has been from people who are from native cultures. I feel that I've been blessed in being able to pass on some of their information.

And do you feel the loneliness in your profession, in the work that you do?

DL: You know, I don't actually feel lonely. Maybe that has to do with the fact that I feel more and more part of all of life. The more you feel that the mountains rise inside of you, that the rivers run through you, and that you are part of all things, the less lonely you feel.

Certainly, when I was younger, I felt lonely, but I don't anymore. I spend a lot of time on my own . . . When I am alone in nature, I never feel lonely.

Once again, it feels as if we have just begun this interview, but Jessica's watch tells us that it has gone on for some hours already. It's nearing time for us to leave.

What is love, Denise?

DL: (With no hesitation) For me, love is the unconditional acceptance of the one you are with. To accept someone for who he is, without judgment, without conditions—that is love.

And in a more universal way, love isn't just for a person or thing; it's a state of being. When I had my near-death experience, I felt this depth of love. But it was different from what I usually experience, because here on Earth, there is always a feeling of separation. "I love you," for example, means that there is a subject projecting a feeling toward an object. During my near-death experience, I felt no separation. It was an experience of unity and oneness. For me, in the universal sense, love is the true depth of unity. So when you love nature, it is about a union, a deep connection that resides in your soul. And all boundaries of separation disappear.

Thank you so much, Denise!

DL: Thank you! You have touched my heart.

A couple of months after our interview, Denise made an important decision: She gave up teaching feng shui, something she has done for the last thirty years. Although she says it was a tough choice, she knows she is doing the right thing.

Instead, she will start what she calls a "soul-coaching" training program that will teach participants how to truly listen to the voice of the soul. She will also demonstrate how we can discover our soul needs and then create a life in accordance with them. In her words, she is looking for people who want to "listen to the yearnings of the soul and help others do the same." She says, "It is needed and it is the time."

Works by Denise Linn

Altars. New York: Ballantine, 1999.

Feng Shui for the Soul. Carlsbad, Calif.: Hay House, 2000.

The Hidden Power of Dreams. New York: Ballantine, 1997.

Pocketful of Dreams. London: Piatkus, 1988.

Quest: Journey to the Centre of Your Soul (with Meadow Linn). London: Rider, 1997.

Sacred Legacies. New York: Ballantine, 1999.

Sacred Space. New York: Ballantine, 1996.

The Secret Language of Signs. New York: Ballantine, 1996.

Secrets and Mysteries. Carlsbad, Calif.: Hay House, 2002.

Signposts. London: Rider, 1996.

Space Clearing A–Z. Carlsbad, Calif.: Hay House, 2001.

HANS ANDEWEG

We now know how to destroy our planet, but we can also learn to nurse her back to health.

In 1992 biologist Hans Andeweg moved to Cologne, Germany, to work at the Institute for Resonance Therapy (IRT). Both the visible and invisible words are very real to him. Because of this, Hans has the courage to venture outside familiar territory. After leaving this position in 1998, he wrote *In resonantie met de natuur* [Resonance with Nature] and subsequently, in 1999, he and his wife, Rijk, introduced a new resonance therapy method: *ecotherapy. Eco* or *oeco* comes from the Greek word *oikos,* which means "place of residence." Ecotherapy can be applied wherever the balance between organisms and their place of residence has been disrupted—houses, businesses, fields, and forests. This vast scope encompasses abstract systems as well: In addition to working with forests and farms, Hans and his wife serve foundations and festivals. Once a business, festival, concert, or other human organization reaches the optimal energy level, the organization as a whole benefits. A greater life force means a higher measure of self-organization. To begin work with this new method, Hans and Rijk founded the ECO*therapy* Center, a center for teaching and for practicing the healing of organizations and the Earth as a whole. In addition to operating in the Netherlands and Germany, he has ongoing projects in Belgium, France, England, Scotland, and Costa Rica.

Though Hans lives and works in Germany, he teaches courses in the Netherlands, which is where we meet for our conversation. He is a tall young man with delicate features, a shaved head, and an athletic build. On first impression, he seems shy but nevertheless self-confident.

At the beginning of our conversation, he tells us that he discovered during his time living in Germany that his roots are truly in the Netherlands. He thinks and operates internationally but remains Dutch at heart, so he is currently looking for an inspiring site in the Netherlands for the ECO*therapy* Center: "I believe that the Netherlands is the best place for our initiative to thrive. We are forging ahead. Jumping into a rapid current and letting myself be carried along on the wave of nations is very tempting. Still, I will bide my time, as the path itself is my objective. I prefer moving in small, deliberate paces to taking a great leap to the dazzling castle in the air and falling flat on my face."

Hans gathers people from all walks of life at his courses. In recent years, his work has attracted a motley group of farmers, psychologists, managers, people from the agricultural University of Wageningen, environmentalists, people involved in forestry, homeopaths, and therapists. This interdisciplinary contact has proved inspiring for the participants. Everyone can learn from each other, especially about new, unfamiliar subjects—such as observing the aura of the trees—that might still make some feel insecure. Sharing energetic perceptions under the motto "Being crazy together is only normal" makes people more confident.

Based on his book, Hans and his wife have designed a four-year program for ecotherapists. Completion of the program qualifies participants to practice independently. During their first year, participants master energetic work—observing, interpreting, and improving the aura of plants and trees. They also learn how to be at ease with themselves and how to open up to or close themselves off from nature. In their second year, participants start individual projects, such as healing their home or farm. During the third and fourth years, they team up with other participants on projects for the benefit of others and may restore the equilibrium of homes, gardens, farms and country estates, theaters, or—during the final stage of the program—a more abstract entity, such as a festival.

Hans and Rijk closely supervise the aspiring ecotherapists during these projects, primarily to teach them that "less is more." Hans says, "The secret lies in simplicity. We operate with major forces, which we teach our students to handle in a sensitive, inspired, and controlled manner. People who work at healing nature often encounter themselves as well. That is why we recommend attending an intuitive development course as a foundation."

During my own courses, I have noticed that some people feel pity for the Earth because of the severe abuse she suffers. But this is merely one possible reaction. I do not believe that the Earth can think in terms of being a victim, though we might if we were in her position. This distinction is essential; it provides the clarity that allows us to avert many problems. Hans emphasizes the importance of remaining neutral and distinguishing our own difficulties from those of others or of the Earth. We also need to be aware of the images of ourselves that confront us. In his words, "A step outside requires a step inside. To ascend, you need to take a step down."

In many respects, Hans's work resembles my field of involvement in recent years. Although he may use different instruments and designs, I am fascinated to meet a kindred spirit whose daily work involves being receptive to subtle energies and the essence behind matter.

What is nature?

Hans Andeweg: (Laughing) Wow! Anything that is not culture and not man-made. The word *culture* comes from *colere* in Latin, which means "caring for, working for, and refining the Earth." Basically, this means spiritual elevation and material refinement of the Earth. The word *agriculture* conveys the idea.

Where is the limit? Can a car be considered "culture"?

HA: Yes, because a car is man-made. A car is refined matter, matter that has been detached from its natural context and subsequently kneaded, sculpted, and molded into a new format. The concept confused me for a long time, but I understand now that it is a refinement. A sculpture is also *agricultura*. The marble has been refined to a higher level.

Is philosophy the human quest for nature?

HA: In some respects, it is. It concerns primarily our search for our internal nature, which relates very closely to nature around us. In philosophy, we think about thoughts. Descartes said: "I think, therefore I am." We base our thoughts on perceptions. My perceptions and my reflections on the subject, therefore, ultimately determine who I am. These perceptions are derived first of all from nature. Thinking about nature enhances my self-awareness. Nature nurtures us in both material and spiritual respects. We have many reasons to be grateful to nature. By cultivating nature and thereby elevating it, we give nature something in return.

In the past few centuries, the importance of this process has diminished, and we have started to oppose nature. We have felt subject to the forces of nature for many centuries, until the English philosopher Francis Bacon said that if we put nature on the rack and forced her to divulge her secrets, we could control her. That was how we subjugated nature. We have balanced our forces against those of nature ever since. Refinement has become synonymous with manipulation. It is no longer agricultura, where you listen and act according to the intrinsic value of nature. We have become far more concerned with what *we* want.

What led you to study biology?

HA: I chose biology because I was searching for nature. The instruction I received, however, related to mathematical statistics and physical chemistry, little atoms I had to stick together. That is how science depicts nature. But that is dead nature. Such matter is part of nature but is not the essential part. If we are unable to contact our own essence, I firmly believe that we cannot connect with the essence of nature, either.

Your ideas sound similar to those of Arne Naess . . . How would you characterize your bond with nature?

HA: I feel it is very close. Often, however, I digress and need to extricate myself from my mental frame of reference and return to my own nature, to be spontaneous and natural, to reestablish contact with the nature around me. Getting in touch with something requires releasing yourself from obligations first—discarding all your obligations, especially the "eternal must." I find this to be difficult. I swim a lot. As an element, water is important for me to let go of my obligations. Once I let go, I can be very satisfied with small things, such as discovering violets growing in a forest and enjoying the little robin that comes to drink from our pond every day. Little things make me happy then.

How do you learn from nature?

HA: By watching, listening, and letting myself be amazed. That is the basic disposition. Be open and inquisitive. Say hello to nature and ask, "How are you?" Then nature will reveal herself to you. Many technical instruments have been copied from nature. The principles underlying helicopters and ordinary Velcro tape come from insects. We humans often believe we have supreme wisdom, which is why we think we know what is right for others and for nature. But if you ask, you will find that nature will tell you exactly what she needs, and that nature demands surprisingly little.

Could you tell me about resonance therapy?

HA: I have spent a long time developing resonance therapy. This technique provides for widespread and remote healing of forests through computers and radionic devices. It took ten years to develop. We treated more than ten thousand hectares [24,700 acres] of forests. We achieved results almost everywhere! I became keenly aware of how to heal on a massive scale with relatively simple devices. This is wonderful! Such an experience is impressive and has moved me deeply. I believe that as humans, we have received enormous opportunities to heal the Earth. We now know how to destroy our planet, but we can also learn to nurse her back to health.

I thought this was important to share at the time but did not know just how important. After I left the Institute for Resonance Therapy in 1998, I wrote a book to depict a course of action for others based on my own experiences. I call it the course of green fingers toward resonance therapy. At the time, I did not yet clearly envisage ecotherapy.

He did, however, envisage turning over the responsibility to the client and wants to teach her or him to do the work. If the job proves too difficult, Hans obviously does not mind helping. Ultimately, however, clients need to manage on their own. Where Hans departed from resonance therapy was in the principle that clients would pay and simply have the job done for them. Clients were passive throughout the healing process. He deplored that those with a sick forest for which they were responsible did not heal it themselves. Hans considers the situation comparable to an ill person holding the doctor responsible for her or his recovery. It very well may be that Hans is able to nurse the forest back to health; however, unless the principal (the client) personally learns to do this and starts to understand the cause of the illness, Hans is convinced that the forest will relapse.

HA: Resonance therapy projects often took three or four years. Nowadays, we work with somebody for a year or two, no more. Our operations cover a massive scale, involving ecosystems of about a thousand hectares [2,470 acres] in places such as Costa Rica, an area of five thousand hectares [12,350 acres] in Scotland, and a national park in Yorkshire. Finding out who runs a project is essential. It is not always the principal. Who is the major cog in the wheel?

The principal or manager infuses the entire ecosystem or corporate system with his or her attention and energy. He can make or break, heal or divide, also in spiritual respects. There is often a patron or patroness. This, by the way, is a beautiful word. These people are often far more capable than they imagine of running, coaching, and inspiring ecosystems or corporate systems . . .

We work toward this goal together. I believe that we are largely responsible for creating our own future. Events are brought about by us or simply happen. They may favor us or be deeply disappointing. If you feel good about your business, you are more likely to do well. This perspective also allows us to show that many social problems in a system arise from long-standing patterns that have taken on a life of their own. They may be events from the past that were very painful, like a huge row. Such incidents repeat themselves. This principle underlies Rupert Sheldrake's work with morphic fields: If something has happened once, it is more likely to happen again. This likelihood increases with each recurrence and leads to habits. Ultimately, the pattern becomes a blueprint in the morphic field. In fractal mathematics, we call this an *attractor*. In esotericism, we call it a *being*.

Companies each have their own being, with partial beings designed to perpetuate themselves. This allows a culture of hatred and envy to take off and acquire a life of its own. Such patterns cause constant chaos or disagreement. Many tensions within companies result from this situation. Often, the staff members are not at fault. Rather, the cause is suprapersonal. The pattern lures people toward a certain type of conduct. Choosing a scapegoat is not the answer. This perspective makes teaming up to find a solution much easier.

So if you locate a problem in a forest or a company and the management and staff start blaming each other, you might say something like "You are not the source of the problem, but rather an old, lingering fear is the problem, or some other negative experience is the problem"? Is it as though a morphic field gathers around the site or the company?

HA: Precisely.

I explain that while teaching a course in France I entered deep meditation and sensed a thick layer of earth containing no trace of energy about ten meters [thirty feet] below the surface of the soil. It felt more like concrete than earth. Beneath this was a dormant layer, that stored all information about the earth there. This layer was dark, mysterious, and vibrant. Contacting and designating it as being associated with the group helped course participants understand that this layer had something to do with our inner selves. Focusing more on this, we found that the amorphous earth layer was actually covering the layer below it, blocking out all the old yet vibrant and precious information about the earth there. During the week of the course, that area shifted. The students became wonderfully lively, while the soil radiated a vital vibrance. This layer in the earth—and, by extension, the students themselves—was slowly awakening. The owner of the old château later explained that the entire sewage system had been clogged, and that, as a result, their business had had a very difficult start. They were grateful that we had worked with this deep layer in the earth. It was beautiful to see once again how the cleansing of humankind and the Earth were compatible.

HA: This is familiar territory to us. I know that some of these layers have been there for a very long period. Transforming and dissolving them usually takes time, like peeling an onion. Compassion and love are essential. At the same time, you need to be alert and adroit, because a lot of negativity may ensue. If you approach an aggressive dog in a kennel, for example, you need to be very cautious to gain his trust. One step at a time you come closer to the core, the reason why he became the way he is. Although you may do this with love, you always need to be watchful. That is how ecotherapy works.

Doing this work with several people makes for support and complementary input. These days, a great many systems are no longer whole. Disintegration is widespread and keeps accelerating. What makes this work difficult is that you can no longer address the essence of a forest. You no longer have a conversation partner. Instead, you have to enter the chaos to search for all parts and make them whole again. That is healing in the true sense of the word. Jesus once said, "If two people gather in my name, I am present." That is one of the reasons why ecotherapists work together. Together, seeing the forest through the trees becomes easier.

How do you approach your work?

HA: We establish contact with an antenna, a "resonator," as Sheldrake calls it: a map or an aerial photo of the forest where we are working. Congruence with the actual forest is very important. With my hand "chakra" I observe the energy of the photograph. Next, I use my intuition to plug in to the morphic field of the forest with my resonator. I make myself neutral, which means that I extinguish everything that concerns me and connect entirely with the morphic field. That allows me to communicate with the forest. I greet the forest being and ask how it is doing. I ask the forest its name. Next come a series of feelings and images about the condition of the forest. Here, too, the first impression is the best, although the problem is how to interpret these impressions, because the forest is such a complex entity. That is why the next step is an energetic analysis in which we use different energetic scales to quantify the condition of the forest. This enables a comparison with other forests and keeps us more informed about our sick forest.

Afterward, we ask the forest how we can help and we test the equilibrium. Each ecotherapist will use her or his own energetic "home remedies," as we call them. These remedies provide the forest with new information to solve a problem. We call these remedies *informers*—for example, a color or a Bach blossom. A *transformer,* on the other hand, would be a mantra, a prayer, or an image of a grain circle. These are energetic keys that open certain acupuncture spots. We offer these items to the forest. Each one has a different effect. The forest's awareness of its needs is fascinating. We do not force anything on it. Moreover, I offer only things that I know how to use because I have internalized them. I could never work with symbols or grain circles I did not know.

How do you involve the principal or manager?

HA: First, we teach her or him to create internal calm. We take our time and do not necessarily start with meditation. Have a cup of coffee or whatever. Reflecting about your inner self is what matters most. Absorb the overall situation. In your mind, stroll around the parameters of your business, or fly across your forest. If that is difficult, take the map and connect with it. Or run your finger along the edge of the forest image to get a sense of it. The objective is to become aware of the entity as a whole and to develop a sense of everything that lives there. Next, principals or managers repeat and visualize their affirmations, or so-called positive objectives. Instead of "There are no illnesses or infestations in my forest," they will say, "My forest is vital and healthy." They make the impression as lively as possible. The moment the principals do this, we perform the equilibrium. We set a time, so that we are all in synchrony at that moment. We supply the energy to move full speed ahead. The captain uses his affirmations to set the course. Synchronicity and consistent repetition of information are highly effective over time.

Do you train the patron, principal, or manager as a healer?

HA: Yes, in fact, I do, but that is only the first step. We also arrange workshops and courses. In the end, they need to carry on by themselves.

I tell Hans that many people label this work "out there," even though it is actually down-to-earth and resonates with the multisensory person, someone who is alert, with all senses active and open. Such people can talk with the sun and all life-forms, as we will hear in our conversation with Emoto, who describes his theory of vibration, resonance, and sojisho.

Do you train people specifically to heal the Earth and its systems?

HA: Yes, absolutely. I see and feel the Earth as a great living organism. Lovelock has written wonderful things about that in his theory about Gaia. You will find it in Sheldrake's work as well. He believes that all systems have a certain measure of awareness. Aware systems can exchange information, which makes for communication. I can communicate with Mother Earth this way as well. The problem is that we need to be on the same wavelength and to speak a common language. This is the hard part. Mother Earth understands more about mathematics than Dutch!

The health of our planet worries me deeply. I also notice that the Earth, and humankind with her, has become involved in a major transformation process. We hope to be of service during this period of sweeping change. Ecotherapy has a vast range of uses.

Do you consider your work to be scientific?

HA: Possibly. Our work is based on intersubjectivity. It is not objective, because we do not use instruments. People can, however, learn to observe auras in a manner that renders their energetic perceptions identical. In their second year of training to become ecotherapists, students work in pairs to observe the aura of ten trees. First, they need to agree on the extent of the trunk aura. Next, they harmonize their observations with those of the other group participants. This takes practice. It is like the annual vitality study of trees, where pairs of researchers receive international training to perform comparable visual observations of, for example, oak trees. This is a foundation for science.

I am a scientist and have spent years researching the effects of resonance therapy with other scientists. Our results keep improving—improved vitality, more plants, more animals in the forest. To my amazement, however, I have noticed that the more results we achieve, the more people have distanced themselves from us. They have become increasingly frightened. They simply cannot believe—resonance therapy, ecotherapy—that as people, we have been able to heal the Earth and its systems on a massive scale. Isn't that too good to be true?

What makes science so skeptical?

HA: Fear—surprisingly, not rational, scientific arguments but something entirely subconscious and emotional: fear of self-preservation, fear of being the odd person out in the scientific fold. Fear of standing out, because that would be professional suicide. By definition, science should continuously explore frontiers and be willing to extend them. But that is not happening, and that is interesting. The paradox is that science is confined by its own research drive. The more a certain mindset or manner of thought is repeated, the more defined is the form that emerges. Science and practicing scientists have restrained themselves through their approach. They are trapped in a fixed, self-perpetuating manner of thought. The essence of science protects rather than extends its frontiers through all different modes of fear. Transcending the borders and stepping out of the mold takes courage.

I admire scientists, such as you and Rupert Sheldrake, who step out of the mold and go farther.

HA: The sense of freedom you regain is the most beautiful part. All of a sudden, you have the freedom to broaden your conceptual scope. In fact, science is not just a lark; it is a serious business. I enjoy life more now, and that is quite an achievement, given that I am fairly serious by nature. I am inclined to feel responsible for all kinds of things.

Everything we are talking about concerns unity. Do you experience such unity in your work and in your daily life?

HA: (Smiling) Sometimes life is too much for me. In that respect, I feel almost *too* responsible at times. Earth really is my planet; I am very fond of her and have felt at home here for a very long time. In this life, I have leaned on different pillars: I have done a lot of music and cabaret. I have always had those two pillars. On the one hand, there has been music; on the other hand, science. Resonance therapy has enabled me to practice both at once. I have always loved telling jokes on stage. This work [resonance therapy] is more serious. There is happiness but less laughter. I derive such happiness from the sense of being united with nature.

Everything is becoming complicated at the moment. These are exciting times. The attacks in the United States on September 11, 2001, were a major blow. Fear seems to have ruled the world since then. I get the impression that forces are at work to smother the Earth in a thick layer of carbon dioxide and leave people to simmer in their own broth. Polarization and confrontation are also factors, because they confirm the old, rigid forms. If you are trying to block transformation and the transition to a new era, this is the way to do it.

Overall, the circumstances of the attacks have turned out to be far more complex than the media would have us believe. Quite honestly, I find global warming to be a greater danger than all the terrorists of bin Laden and Saddam Hussein combined. Global warming is melting the polar caps and the glaciers, which leads to more river water running into the sea. Research by the Max Planck Institute in Germany has revealed that this process can bring the warm gulf stream to an abrupt halt. The climate in western Europe will become like the one in Siberia. Once that happens, all the oil in the world will not be enough to keep us warm.

Even in this case, we should not let ourselves be driven by fear. I often experience fear as a thought that has become an independent being. There are indeed beings of fear who delight in terrorizing us. They feed on all those energetic fear frequencies. Finding inner tranquillity helps. You will notice how your thoughts will change and become more positive. There is always a way toward a future of hope.

I notice how moved you are by the Earth, how much you love her. Is the feeling mostly love or pain?

HA: Both. Pain is part of love. The agony of love is part of being human. We journey to the very depth of the matter and go through a kind of death with Earth, to be resurrected in the light of a new era. Earth knows what is happening. She was made aware by Christ two thousand years ago. I am at peace with the events on this planet and do what I can. I always tell myself, "Keep within your means. Do not bite off more than you can chew. Set goals that you can attain without overextending yourself." My aim is to bring something into the world that will eventually continue without me. I also hope I will have more free time to make music. I often feel as though I am a kind of pacesetter, a pioneer. The idealism vanished a long time ago; we all have to earn our keep. In this world, you need to be businesslike and pragmatic in addition to being idealistic.

My own experience is that the more cheerful you are as a healer, the stronger your energy, while the more serious you become, the heavier your energy becomes. So music and fun will only improve your work!

HA: Tell me about it. There are plenty of people who sing well and write nice songs, but I have yet to meet any who work with ecotherapy. (Grinning, with disarming honesty) That is why I am focusing on this field . . .

Could you heal polluted patches on Earth through clean places on Earth?

HA: Absolutely, but you need to be careful. Establishing an energetic connection between them will channel the energy from one site to the other. But there will always be a measure of reciprocity that will make the bad site influence the good one. In all cases, you need to request permission from the parties that control both sites.

Suppose you connect a clean portion of land that I have the privilege of managing in South Africa to an unclean portion of land in Costa Rica. Is that possible, and what will happen?

HA: There are many ways to do this. What matters most to me is that the energetic pollution from Costa Rica does not reach South Africa.

The frequency of your land should be high enough and the vibration sufficiently rapid to ensure that the slow, diseased vibration from the land in Costa Rica does not take root there. The moment that I jot down the names of both places and draw a line between them on paper, I have connected their energies. The arrangement is like the line of communication you and I are having right now, which is always reciprocal. Another way is to use the power spots in your area—to depict them in a photograph or symbol and in this way work with parts of the whole system. That is better. I would place the photograph on the map of the forest in Costa Rica. This connects the two morphic fields with each other and channels energy and information back and forth.

You work a great deal with disease, with what is not healed. How do you keep healthy?

HA: I meditate daily, get lots of exercise, stay in good physical shape, and relax now and then. Enjoying life and having fun are very important! (Grinning) I enjoy a good whiskey, and we go to the movies a lot.

The image of the two triangles that you describe in your book seems to be filled with hope. I interpret it to mean that you do not place things beneath you or above you. If you are truly complete, though, you will be open to anything and will remove the "screens" from your heart. This also alleviates fear of people and of situations. You overcome fear of them through your tremendous love. Can you tell me about that?

HA: Our heart is our center. One triangle connects the heart with the spiritual impulse, the cosmic. The other triangle represents matter and runs downward from the heart. Everything begins with a meeting of cosmic spirits and Earth matter. This is where life is created. First, there is a speck, the seed. This speck opens, and the seed evolves and becomes a revelation. The life processes increase as spirit and matter permeate each other. A species emerges. This is the course toward healing and becoming whole. Becoming holy is the next step.

I feel the connection between heaven and Earth everywhere. To me, healing means bringing heaven and Earth back together. This wholeness gives rise to a center field. In people, this field is our heart, the middle area: head in the sky, our feet on the ground, and living from the center with our heart and soul. It is our way to restore heaven on Earth. Rather than trying to reach the seventh heaven as soon as possible, we need to create the eighth heaven here on Earth. People in search of enlightenment are eager to get in touch with higher energies but often forget the Earth. Before they know it, they have gone too far. Our spirit is faster than light, but our bodies need time to integrate higher energies. It is like wanting to read university textbooks in primary school. You develop a sort of hydrocephalus while you are aiming for the middle ground.

In one of your articles you describe the rage and dismay of a forest. I do not see how this is possible. We humans are the only creatures that have rational thought. Therefore, we judge. The other species do not judge.

HA: Emotions are different from rational thought. As a scientist, I would not have formulated my statement the way it was phrased in the article. We tend to apply the conceptual framework of humans to other species. This is wrong.

In my view, the healing quality of nature derives from the nonjudgmental aspect of the natural. That is why we feel free to be ourselves in nature, which many people find deeply soothing and inspiring.

HA: Observing a site and devising common concepts with a group of people during a course is fascinating.

Would rage be one of the concepts you encounter?

HA: Well, the group would certainly come in touch with sadness when a site feels sad, depressed. I have devised scales in my book, based in part on Wilhelm Reich's energetic concepts. He distinguishes *orgon,* the flowing, healthy life energy, from *dor,* the dead, blocked energy. *Oranur* symbolizes stressed, hectic energy and has an almost hyperactive quality. It is very sharp. You feel it tingle. Dor is very heavy and sticky—like a heavy, gooey energy. This enables me to avoid a debate as to whether or not it is rage. Dowsing the percentage of dor and oranur at a spot with the help of intuition and a pendulum is easier than having a discussion about feelings.

I have my course participants experience what fifty percent oranur feels like. Individual perceptions vary. Those feelings and images are subjective. After extensive practice, the percentage perceived becomes intersubjective. In addition to dowsing the energy, I always ask course participants to tell me their feelings and impressions about a spot. They learn from each other and practice being nonjudgmental. We are quick to speak in terms of positive and negative energies, but withholding judgment as long as possible is important. Those interested in conducting healing should refrain from judging and dividing.

What is soul and what is spirit? Everybody interprets the concepts differently. One person will say the soul is eternal; another refers to spirit in this way. What do you think?

HA: I have avoided referring to either one. While I consider the existence of a soul and a spirit beyond dispute, these concepts get so confused that I try not to use either term.

***But you liked the concept of* soul.**

HA: The concept of the soul concept appeals to me. I associate it with emotion and solidarity. The *soul* concept seems like something impassioned that embodies the divine breath and is alive. *Spirit* sounds more aloof and abstract, like something that concerns information. We base much of our work on the principles *life energy plus information equals formative strength* and *formative strength plus matter equals form or species.* These forms and this formative strength exist throughout nature.

Would you describe what you see in Masaru Emoto's work as the formative strength in the water?

HA: Yes. He uses tap water from Tokyo, New York, and several other major cities, as well as water from running creeks and holy springs. You can see a beautiful crystal emerging from the water of a holy spring, while the one from tap water is thoroughly unattractive. This seems logical because the holy spring connects with higher energies and more-complex information than does tap water and thus generates a complex, richly structured shape expressing this formative strength in water. The shape is fascinating; it is the center of activity. It is a *holon,* a whole with an interior and an exterior. This may explain my preference for the term *soul.* The shape arises from life energy, spirit, and matter. The spirit inspires the shape and gives it substance. I am just improvising now. (Smiling) It sounds good, but I may feel differently tomorrow.

So does the soul continue?

HA: The soul experiences. It goes through life and gathers experiences. After death, the soul is confronted with itself and with its emotions and cleanses itself of them. The soul passes through fire to this end. While the process is cathartic, it is more of an inner fire than a fire of hell. After purging its emotions and discarding its emotional ties to earthly life, the soul moves on to the spiritual world. It may be reincarnated on Earth to gather new experiences or come to terms with old pain.

Sheldrake talks frequently about terms and their meanings. He previously referred to resonance therapy as "sympathetic magic."

HA: In 1992, we discussed our ideas and activities with him. Hearing him use the definition "sympathetic magic" was quite disappointing at the time, as if resonance therapy were inexplicable as a method. In fact, we demonstrated how his theory of morphic fields was applicable in practice. He seemed to feel we were aiming too high and said so. He had enough trouble explaining to people that the morphic fields of blue tits resonated through form resonance. He did not even want to think of explaining to scientists that a photograph of a blue tit resonated with the blue tit field, that humans could use such a photograph to contact the field and may subsequently request information intuitively. But that was a long time ago. I have tremendous respect for Sheldrake. Who knows? We might revisit the subject some day.

What is love?

HA: Love is everything and nothing. It is the great alchemy in this universe, the all-encompassing power that connects everything with everything. Love completes the beginning and the end. You could also call it God, the Great Love. God is the creative force in this universe that lets everything expand and holds everything together. God has created this universe and all others. God's spark connects us with our deepest essence. Willem Kloos expresses it beautifully in one of his poems: "I am a God in the depth of my thoughts." That is the truth to me. We are God's children and are born from love. It is not a romantic love, where you think somebody is beautiful and sweet. Other aspects come into play as well, from crystal clear to tough and merciless. Loving somebody means yes as well as no. Many people find that difficult to accept. I may love nature. Or my computer, or my car. I may even perceive them as beings. It is a different expression of the same. I think we are complete as humans only once we have experienced and lived the Great Love in all its facets. That may take awhile.

We need to manage together. That is the only way. I was raised as a humanist, along the lines of: You live, you die, and that's all. Even as a child, I sensed there was more to our existence, but I did not get stuck with some kind of religion. I am very pleased about that.

Can you associate what you are doing now with the way you felt as a child?

HA: Absolutely. I have observed that many young children are very open and perceive things invisible to the ordinary eye. Parents in our culture rarely know how to respond. We suppress a lot. I had all kinds of strange and scary experiences as a child. My parents would say, "There are no ghosts." They meant well. In fact, their intentions were very good. They did not want me to be afraid.

For example, I always had a great big lion in my room. It was simply there, and it looked creepy. It was so big. I had a little teddy bear, but that lion was huge and not really cuddly. When I think back, I suppose the lion was there to protect me. I think it was a kind of totem animal. Perhaps it had to do with a kind of power within me that was not yet incarnate at the time. Talking about things like that with my family was difficult.

Did you tell your parents?

HA: I think I did, but I was not an easy child. I had many weird ideas, and my parents would wonder what was true, what was imagined, and what was feigned.

His family's attitude has changed a great deal. His mother worked at a library, and when Hans lived in the Netherlands, she often took out books for him. Eventually, she started reading the books that fascinated her son. His father produces all kinds of materials for his courses, and his sister is in her second year of the program.

HA: (Smiling with visible pride) It has become a family business.

Does that mean you do not feel lonely? Not even as a pioneer?

HA: Definitely not. I enjoy talking about it. The Netherlands has changed a great deal in the past two decades. Nowadays, I find that people are more open and less resistant. I started as a researcher in organic farming at the Louis Bolk Institute in Driebergen. Around 1988, reports appeared on the news and on the front page of the *NRC Handelsblad* about a huge study I was conducting on homeopathic drugs. I am used to being in the public eye. One minute I am being praised to the skies, the next I am maligned. You learn to live with it.

Sometimes it is hard to take. In 1992, for example, the IRT commissioned a three-year preliminary study on the impact of resonance therapy on Dutch forests. The results were promising. There were enough positive findings to justify a vast, in-depth scientific study. The Ministries of Housing; Spatial Planning and Environment; and Agriculture, Nature Management, and Fisheries allocated financial support. The Vereniging Natuurmonumenten [Association for the Preservation of Nature] agreed to make nature reserves available, and the University of Wageningen was to conduct the research. We received widespread media coverage, and the press reviews were largely favorable.

Everything seemed settled until the counteroffensive got under way. One extended negative article appeared in one of the national dailies, and the damage was done. We lost our government support overnight. Notwithstanding all the favorable media coverage, this article was considered far more important. Even though we proved that it was riddled with inaccuracies, the commission for the study was canceled. The government felt that public sentiment had changed. The article was published all over the world and turned opinions against us. I had tried so hard to conduct good research about my subject and was undermined in a manner that was underhanded and unfair. It hurt. I realized that we live in an unjust world. I was devastated. As I said, though, I have come to terms with it and have learned from such experiences.

In many respects, Rijk and I have learned the hard way. At the Institute for Resonance Therapy, we were thrown in beyond our depth without a lifeguard or even a life jacket. All we were told was "You can do it." We had to sink or swim. We swam, and, as you can see, we still are [swimming].

As our work progresses, we encounter more opposition. We have to deal with it. The world is filled with jealous forces that seek to block innovation. This is the reality, but we are not frightened. I have deep faith in humankind and in life. Things always work out. The morphic field of innovation keeps growing bigger and stronger. That is why somebody like Sheldrake is wise to continue to publish books. This approach increases popular support for his work.

We also need to keep working on ourselves and to manage our own energy wisely. We have a central identity and a peripheral identity. Central identity is the way you are as you sit here. Peripheral identity is the part of you that lives among others. If you write a book that others read, something will happen. Your peripheral identity is operating at that moment. As more people read your work, they will reach out to you. This forces you to move farther toward your own center to recharge your energies before nurturing others. Emoto is now publicizing his work and is consequently expanding his peripheral identity, but is also making himself more vulnerable. He will have to work to maintain his energy level.

Peripheral identity is congruent with who you are . . . You provide a beautiful, clear explanation.

HA: This is why some rock stars stick around while others soon suffer from burnout. [Some] exhaust themselves with their first hit. They literally evaporate into all their fans. All that remains is an empty shell, an empty feeling, and, in most cases, a person suffering from severe depression. I admire artists like Madonna and the Rolling Stones—true stars who are still shining after all these years.

Is there anything you would like to add to our conversation before we close?

HA: We are here to evolve to a higher level with the Earth, and in this adventure, too, the process is the objective.

Works by Hans Andeweg

In resonantie met de natuur [In Resonance with Nature]. Utrecht: Kosmos-Z and K, 1999.

Succes: lessen in Maatschappijleer [Success: Lessons in Social Studies]. Kampen: Kok Educatief, 1986.

MASARU EMOTO

I would like to say that nature in this universe consists of the concepts of love and gratitude.

I meet Masaru Emoto in an office in Amsterdam. He is in the Netherlands for the wedding of his daughter, who studies at Leiden University. On a hot day, we sit at a large table with many bottles of water. Masaru has come with his wife and a secretary. Because I speak no Japanese and Masaru speaks little English, I have arranged for an interpreter.

We start with smiles and thanks, and then we begin our conversation with the subject of the element water. Masaru is known worldwide for his experiments focusing on the influence our thoughts have on water. In 1994, he developed a method to take photographs of polluted water that had been frozen to a temperature of –5 degrees Celsius. After thawing and inserting this water into test tubes labeled with the title of a musical piece by Chopin or a name or a word, he freezes the water again and takes a second photograph. In this second photo, the water has transformed into a beautiful crystal from the formless msss it had been in the first image. In this way, he has succeeded in presenting "the invisible world in a visible way."

With work like this, Masaru shows the world the power of our thoughts and how they influence our surroundings. His work is not scientific, but rather delivers the heart-filled message that we can heal what we have destroyed—that with our focused attention, we can make miracles happen. The message is clear and strong.

The beautiful crystals he has photographed are awe-inspiring. My own experience that we can heal people, the Earth, and animals with focused attention opens my mind to his loving work. I am also aware that each person's thoughts, as well as actions, have a direct influence on our surroundings, which makes us responsible for the consequences of how we regard the world around us and act upon it. When my children were still small and living with me, they were always happy when I myself was bubbling with joy. Our pets also feel our grief, anger, or happiness and react to it. For instance, the cat jumps on my lap when I meditate. Why, then, would not water react to our moods and thoughts?

We talk for hours and forget the time until our empty stomachs start to protest and we realize it is quite late in the day. The interpreter has worked hard to translate precisely—not an easy task, given our different cultural backgrounds and the different mind-sets, reference points, and perspectives they comprise. Our differences, though, make the exchange even more interesting.

Past our smiles and the differences of language and culture, we connect with what really interests us: the Earth and our role on it. The interpreter is late to our meeting, and we decide to start our conversation without her. We try speaking English.

Do you understand my English? Please stop me when you don't understand.

Masaru Emoto: Now okay. My wife has actually read your book *Dialogue with Nature*.

Indeed, my book was translated into Japanese.

ME: Yes. She liked it.

Mr. Masaru, you've studied subtle energies for fifteen years?

ME: Yes, using a special machine named MRA, Magnetic Resonance

Analyzer. I am a specialized operator of that device and I can check very subtle energies from human bodies—even emotions—or water or plants or anything. I can diagnose or check.

So when a plant is scared of someone, you could measure the vibrations?

ME: Yes.

I have read about that . . . I would like to start out by asking what **hado** ***is. It seems to be the central concept of your books.***

ME: Hado is a subtle energy that has an important frequency, being influenced by emotions or consciousness—human consciousness.

Masaru explains that hado is a pure vibration, like the beginning of energy—the vibration of atoms: electrons, and protons. The core of an atom is always vibrating, he says, which is a tiny, subtle energy. Masaru's theory is that the first energy field is so low that only the consciousness of human beings is able to affect or influence it. He describes hado as our essential energy—the essential energy in all life-forms. He agrees when I suggest that it might be the energy that connects all life. Some might even call it godlike.

It seems our positive or negative thoughts can influence hado in much the same way they affect the changing crystals that Masaru photographs. This suggests our important task in effecting positive change: to think beautiful, optimistic thoughts. But it can work the other way around as well: Anything that is good or beautiful can be influenced by negativity. Ill thoughts can cause a crystal created by good thoughts to disappear. We talk about this in relation to Masaru's work:

But you haven't shown that—a crystal that is beautiful becoming, in the second photograph, something that is a formless shape?

ME: Depending on the photographer, the result is of course different. If a photographer's heart is warm and nice, the pictures are very, very beautiful.

So the photographer also influences the result? If the person taking the photos has bad thoughts, might the crystal dissolve? Can our thoughts influence the crystal that quickly?

ME: Immediately. The crystals are no more than one point five centimeters of ice. I see this as a small universe. If a photographer can see a very, very beautiful landscape in this universe, this universe will be very happy. But if a photographer cannot see a beautiful landscape, if he can see only a bad landscape, this universe will be very unhappy.

What happens then?

ME: The universe can go in any direction. It depends on our consciousness, human consciousness.

Filled with awe by the implications of what Masaru says, and how in a few words he shows us that the future lies in our hands, I venture to the next question, a very difficult one.

Does water feel? Credo Mutwa is convinced it does.

ME: I don't think so. Water doesn't really do anything. Water is only a carrier. Its role is in fact as a mirror, a mirror of human feelings or human consciousness. So if my mind is bad, the water is the mirror of my mind, which means that the water becomes bad. If my mind is good, the water becomes good, because it is a mirror.

Is it also a mirror for the plants, or for the sun, or for animals?

ME: Yes, sure, sure. Actually, my theory and my work are more acceptable to the European people than to the Japanese people . . .

At this point, the interpreter arrives, apologizing for the delay. Masaru, although distracted by the interruption, continues more comfortably.

ME: I believe the role of human beings here on Earth is that of a manager, so that all living creatures can coexist happily. But human beings are using a lot of chemicals, and as a result of this, they are polluting the water. The vibration, or hado, of such chemicals is quite strong. The vibrations, or consciousness, of animals cannot exercise influence on water, unfortunately.

I believe that all nonhuman living species don't pollute or don't think bad thoughts because they can't do these. Perhaps in this way they can influence each other—and water—so much that they can neutralize a bit of human's negative hado.

ME: I think so, too. But there are more than six billion people. And we all think bad and good. Even if all animals have consciousness energy, the total of human consciousness energy far exceeds that of the animals.

Do you agree that we are the only species on Earth that destroys consciously?

ME: We have to accept that.

But then it is of great importance to restore nature, or to maintain pieces of nature that can restore themselves, to get its energy flowing on Earth again.

ME: That is something I worry about. I am very much anxious about that issue. Nature can protect itself only by completely rebuilding again from the destruction. If we keep on going the way we have been, that would be the only way for nature to protect itself.

What is nature to you?

ME: Nature is the phenomenon of circulation and revitalization of love and thanks. This is created by water, which transmits energy [vibration and information].

I tell Masaru about my life in Africa—how I have been in the fortunate position of being able to buy six thousand hectares [about 14,800 acres] of semidesert land in the mountains of South Africa and how I am simply allowing it to restore itself. I've returned some of the native animal species, and the grasses and flowers are coming back. Together with the neighbors, I dream of making a larger nature reserve.

I think it's very important that there are places on Earth like that, where everything in nature can just be. It's also important for us to have places—beautiful places—where we can rest. The Earth needs the energy of such places. Do you agree?

ME: You are exactly right. I believe that as there are more and more protected areas, the human mind will also think more of natural space. I think that then we will be able to avoid the catastrophe on Earth. In my opinion, all the symptoms of this evil lie with the oil industry.

Your work shows that our thoughts can have an affect, which means that every person on Earth is important, that all individuals can change the world with their thoughts, with the choices we all make—simple choices, such as what we drink, where we work, what we buy, what we say about others . . .

ME: Indeed. But there are those who understand this and those who don't understand this kind of thinking.

Yes, but that is also a choice.

ME: So my role is to travel around the world to transmit this message by showing these photos. And when the percentage of people who understand this way of thinking increases, it will help to restore the [Earth's] energy.

I agree completely . . . But I think it's sad that many people on Earth at this moment don't realize they're worthwhile, that they can change things, make a difference. Most people don't feel loved or are insecure or are very angry inside. I think there's a lot of work to do, but I think understanding your work and the importance of thought can heal many people.

ME: I wonder why this kind of discovery was given to me, along with the task of telling the world about it. It is a tremendous thing. It is very necessary, what we have been discussing. It has to be done. In order to save the world, this message has to be transmitted. But why me? I am still mystified by it.

I respect a microbiologist named Teruo Higa. He is associated with a concept called EM [Effective Microorganism technology]. He often says that ten percent of microorganisms are good, ten percent are bad, and eighty percent are opportunistic, siding with neither the good nor the bad. He taught me that the same is true for human society. Changing the awareness of all six billion people on Earth is a huge task. But there is the potential of the ten percent of people who are really conscious and active-minded . . . So, what I would like to achieve is to reach this ten percent, those people who are conscious.

We do not see these people because we do not recognize them. But they exist among six billion people. They number six hundred million people, because that is ten percent of the world population.

What is EM?

ME: Teruo Higa is a professor at Ryukyu University. He is known worldwide today. By using effective bacteria, he has created wonderful soil for agricultural purposes—rice fields and farmland—and has done some miraculous things all over the world.

***The message of the water you photograph seems to be: "Take a look at yourselves." What did you want to say with the title of your book* Messages of Water?**

ME: Water reflects our awareness. That is my title. We cannot really understand if we do not approach with our heart the problems of water.

First of all, let's think about what water is. Seventy percent of our bodies consist of water. Seventy percent of our Earth also consists of water. (Taking an ice cube from his glass) Here is a piece of ice. When you put a piece of ice like this into water, it floats. But nobody knows why it floats. When you place a handkerchief in water, the cloth soaks up water . . . The specific gravity of water is heaviest at four degrees Celsius, but nobody knows why. So far nobody has done a Nobel Prize–winning study on water. [Yet] research on water lies behind all other studies. Contemporary scientists find it most difficult to deal with water. They exclude their heart-consciousness. But nothing about water can be understood unless you deal with the heart of water. Only the religious world has dealt with this. I would like to believe that the origin of the word *crystal* is Christ All.

So water is pure.

ME: That is what I'd like to say, that the crystal is also Christ, and that this is probably why Western society understands this concept.

Do you mean to say it is like hado, in its purest form?

ME: Of course. According to my interpretation, the word Christ means the creator of nature, something great. And its message is the crystal.

Do you think that at the moment a crystal is formed it consists of pure, clean energy—Christ energy?

ME: In the beginning, yes—when the crystal was formed the first time.

Why only in the beginning?

ME: Because the will of human beings has changed it.

I don't understand. Could you explain further?

ME: We are seventy percent water. Without water, no life can start or continue to exist . . . Actually, four and a half years ago, the University of Hawaii, NASA, and Professor Frank of the University of Iowa and leader of the NASA team began studying whether Earth might have been without water originally! The reports from these two institutions state that one hundred tons of water from comets falls to Earth annually—ten million pieces of icy comets. For the last fifteen years, NASA has observed black objects falling to Earth from space. When these approach Earth, they are absorbed. Because they are ice, they melt due to friction when they enter the atmosphere. They turn into vapor, clouds, and then disappear. We can say only that these ice particles come from the universe.

Despite the fact that such significant facts were officially publicized and reported to the whole world by the UP, Reuters, and such, scientists worldwide ignored it.

So through your water crystals you attempt to help us become more aware.

ME: This way of thinking of mine is very religious, spiritual . . . We don't know who we are. We don't know what life is. We don't know where we are from, and we don't know where we go after death. But when I recognized that we are made largely of water, I began to understand. Of course, we come from the universe. And when we die, we will continue our voyage of circulation through the universe.

So we are part of the cycle of the universe as a whole?

ME: Yes.

That is very beautiful.

ME: I realize that it sounds awfully romantic . . . Regardless of whether they are expressed in Japanese, English, or German, the words *love* and *gratitude* cause water to react by forming the most beautiful crystal. Because water is a mirror of various states of nature, I believe that love and gratitude are the principles ruling nature's states.

But if water reacts to an encounter with us, isn't it more than just a mirror? Does water also have a soul?

ME: We could say that water is pure and it has no self. Water itself is pure, selfless love. Let's talk about it.

It has no self?

ME: No self; it is pure and that is love.

With this response, Masaru has already answered my question "What is love?" His explanation moves me and invites me to ponder: Is love self-

less? Matthijs Schouten speaks of opening our hearts to the "other" without prejudice. Loving unconditionally means letting go of our ego, eliminating surface noise, and reaching for clarity.

I explain my thoughts to Masaru: that every species on Earth that has life in it—and water is life—has a consciousness and responds in one way or another. So rather than being a pure, neutral mirror, couldn't water, which becomes a beautiful crystal, have a consciousness that recognizes love?

ME: Probably what I'm going to tell you will answer your question.

First of all, I must reflect on the meaning of *love* and *gratitude*. In order for nature to exist, two types of energy, like male and female, are necessary. In China they are referred to as *yin* and *yang*. I would like to say that nature in this universe consists of the concepts of love and gratitude. I think love is an active energy while gratitude is a passive energy. In terms of the sun and moon, perhaps the sun is love, the active energy. The moon is gratitude, the passive energy. In the case of fire and water, fire is love and water is gratitude.

Water is H_2O. Nobody knows why H_2O makes water. Because fire cannot burn without oxygen (O), O could be considered love. And hydrogen (H) would be gratitude. So two gratitudes and one love together (H_2O) make up the basic composition of water . . . One love and two gratitudes. A metaphor for this may be, for instance, a fish who produces roe—its babies—but lets other fish eat its roe. This is one love and two gratitudes in the *wa* [harmony] of nature. Love is "for me, for me" and gratitude is "for you, for you."

Now, we humans have reduced the composition to one love and one gratitude. That's why this world has become materialistic. The population increased and nature was destroyed. But the essence of water, of nature, is one love and two gratitudes. If we understand it and if we live accordingly, we can quickly return to the way it used to be.

I hope you understand: In this case, I refer to the Japanese conception of love. I am very aware that Westerners have a specific sensitivity about love. But true love consists of one love and two gratitudes. I hope you understand . . .

You are saying that love has a broader meaning than we consider it to have in the West—that love is made up of receiving and letting go again, passing on? This is what I consider to be unconditional love: love that does not want or need to possess or understand or hold . . .

ME: Of course, I think so too . . . Another aspect to the two Hs in water, to the two gratitudes, is that one is a plus and one is a minus. There are two nuances to gratitude: respect and gratitude. That is what I have come to believe.

Then the message of water is love . . . I believe that everything in nature is teaching us, asking us, inviting us, to be open and receive. People are starting to realize that we can receive nature's love . . . we can start to open up. All life-forms communicate with each other, and we are part of this communication. But we have put ourselves in the very center of the universe without listening. While we are learning to receive more love, we must provide a gift to other life-forms through our growing sense of gratitude and respect.

ME: That is indeed the principle of the theory of energy as well. They call it *resonance*. As you know, if we talk or just look at flowers kindly, they will bloom long and beautifully.

I know that there are quite a few people who understand or simply can do those things. They are able to talk to flowers or small insects, such as small butterflies. Although this is not necessarily a pure form of gratitude, I feel part of my role is explaining to people why something like this is possible.

Yes—people don't know this anymore. It is only the old cultures that still do.

ME: I believe there is one fundamental law in this universe: the Law of the Octave, with three aspects—vibration, resonance, and *sojisho*. *Sojisho* means "compatibility." In my own words, this is the do-re-mi theory, or the theory of the musical scale.

When we look at things through these three aspects, it is not at all difficult to understand them. Human beings can naturally sing the musical scale, do-re-mi-fa-sol-la-ti-do. And no matter how high the vibration gets, it is still a repetition of do-re-mi-fa-sol-la-ti-do. There are no other notes.

If we add the tones of the five black keys of the piano, then there will be twelve different tones, or notes, in an octave. Any vibration can be classified into one of these twelve notes. It is said that the human body consists of vibrations that have a range of forty-two tones.

Can we hear these tones?

ME: We have the potential to hear all these tones. However, in reality, it is said that human beings are able to hear frequencies only to twenty thousand hertz. Yet the Law of the Octave is a natural law in this universe. Everything in the universe has its own vibration. Crystals, plants, trees, and so forth have their own vibration.

As humans, we can hear and produce all of the seven or twelve tones through our vocal cords. Considering the Law of the Octave, this means that we are able to resonate and communicate with any frequency in the universe, including that of trees and water. This is the theory of sojisho and Shinto, which is the Way of the Gods, the original pantheistic religion of Japan.

Does this mean that if we open up our hearts, we can "hear" other vibrations?

ME: Indeed. By doing so a human being has the potential to sense and resonate with all vibrations.

Can we—all human beings—hear the plants, the trees, the water?

ME: Without a doubt.

Can we develop that ability, enhance it through practice?

ME: Without a doubt.

That's what I teach people, and Hans Andeweg works with this principle too, as do many others.

ME: Each of us has in himself a world that vibrates. Why do people meditate? Many do not know why they meditate. I was invited to a meditation group called Maharishi. In my meditation group, the wonderful, angel-like love of people's vibrations touched the water of a lake and streamed upward. For many hundreds of years, people went to this water with gratitude. Such feelings continued to be cast onto the water, and even today the lake creates beautiful crystals. After hundreds and thousands of years, it could become holy water. So, water can have individuality—I think that is what you are trying to say, too. But in the end, water reflects only the energy created by people.

How did you come to be interested in water?

ME: Originally because of my business. I myself am an ordinary person. But through my work, I came across a machine called an MRA, magnetic resonance analyzer, which I have mentioned earlier. It was in fact a machine related to homeopathy. It extracted information from patients and transcribed it to homeopathic solutions. Such homeopathic solutions contain alcohol. But homeopathy is prohibited in Japan, or rather not recognized, not even today. To avoid the alcohol, I used micro-cluster water instead.

Individualized information unique to each patient is detected by the machine and imprinted onto the [micro-cluster of] water, and the patient drinks it. I was involved for a long time with this type of therapeutic work, and with research I cured more than ten thousand using this for people with various illnesses.

You are a nature doctor, I understand.

ME: Indeed, I am. People said that it was absolutely impossible that someone could be cured with water alone.

But you add something to the water?

ME: No. Just hado, through pure thought.

I consider myself a magnetic resonance analyzer. Magnetic resonance in fact becomes information. So, when I gave treatments to patients, there were results: Their difficult diseases, such as last-stage cancer, were cured. That they were cured by water was very important, so I wanted to prove it. I wanted to make it visible to people. Day after day, I thought about *how* I could prove it. One day, I read a passage in a book that said that there were no identical snow crystals, and I thought "That's it!" That was nine years ago.

I took thousands of photographs. About two and a half years ago, I thought it was about time that I published them. Then I came to realize that there were tremendous messages in them. I considered that the transformation initiated by such messages was important in terms of curing diseases of each individual patient, so I started giving lectures and began to write.

Was there any personal attraction to water when you were a child or when you were young? Are you water by sign?

ME: I was born on the seaside and I am Cancer—July 22. It seems to have something to do with water. I strongly feel my destiny. My name, Emoto, means "the origin of a big river."

There you are. There is always a spiritual connection to why we do things or why we are here in this lifetime. We can call it obedience to our soul.

ME: If you are interested in making such connections. For the past few years, the top of my head has had a bulge, like an antenna, where it used to be flat. Since I was a child, I have had many scary dreams hundreds of times. If I think back on them now, they clearly have had something to do with the destruction of the Earth. Tsunami, great floods, the ocean splitting open, waves rising several tens of meters high: It was always the same. I have fewer of them now, but I still see them.

Are they becoming fewer because you are doing something with water?

ME: I think so. I hope that I won't have them at all in two or three years' time, because the feelings of love and gratitude increase the quality of water.

Many things can accomplish this increase: Children can simply write on a plastic bottle or on a water-cooler bottle words of love or gratitude. At first in my experiments, I wrote the characters for respect, gratitude, and love on the test tubes containing water, but I couldn't obtain clear crystals. Then I changed the order to "respect, love, and gratitude," with *love* in the center, and clear crystals were formed! It is a great message. The words were my own, but the message is greater.

How does the element water relate to the other elements: fire, air, and earth?

ME: Among air, fire, earth, and water, the role of water and its relation to other things can be explained in this way: Without air, fire does not burn. Without fire, energy [vibration] is not created. Without earth, there is no material to burn. Without water, energy [vibration] is not transmitted. And without water, air cannot come into being. Therefore, water is the starting point of all matter.

Works by Masaru Emoto

Food Science of Hado (Masaru Emoto and Akiko Sugahara Takanawa). Takanawa Publishing Co. Ltd., n.d.
The Human Science of Hado. Business Publishing Co. Ltd., n.d.
Messages from Water, vol. 1. Leiden: Hado Publishing, 1999.
Messages from Water, vol. 2. Leiden: Hado Publishing, 2001.
The Prelude to the Hado Age. Sun Road Publishing, 1999.

Web site: www.hado.net

RUPERT SHELDRAKE

Most people in the Western world . . .
have two attitudes toward nature.
One of them operates from Monday
to Friday . . . the other one . . . comes into
play during weekends and holidays.

The finest sounds of birdsong wake me up. I cast an eye on the clock that tells me it is only 4:30 A.M. I lie there, listening for a while, enjoying the beauty and clarity of the different tunes, as yet undisturbed by man-made sound.

When I open my eyes again, I see the first soft light of a humid summer day coming in through the large open windows. I wanted the birds to wake me because I hate the alarm clock. I am glad they succeeded because today I am to catch an early plane to London to meet Rupert Sheldrake, the scientist credited with bringing ecology to a new level.

On the road to the airport, I come upon masses of cars filled with people already driving to work at six o'clock in the morning. We look like a row of busy ants. What a difference from the quiet of the early-morning traffic in South Africa on the day I went to meet Gareth Patterson in the Knysna woods!

Still, these two men—Rupert Sheldrake in London and Gareth Patterson in South Africa—who live such different lives, have come to the same conclusions: They both love the Earth and live in accordance with their own nature.

The green bus and Underground take us from the airport to Sheldrake's house in Hampstead Heath. A smiling young woman with bright-colored hair sticking out at all angles answers the door. She lets us into a home filled with books—books in the hallway, in all the rooms. It is a warm and welcoming home in which work seems to be constant and important, with the clearing up after it farther down on the list.

After a handshake, Sheldrake leads us straight to a small balcony filled with flowers. He shows us a little aquarium placed on a chair that holds some algae, plants, and an empty flowerpot rising out of the water. He points to the tiny tadpole swimming about. With boyish enthusiasm, he tells us how he released the tadpole that lived in the aquarium before this one. After it became a frog, it came up out of the water onto the flowerpot for its first breath of fresh air, as all of its predecessors had done: "I brought it to the garden and held it there in the open palm of my hand. With one large jump, it disappeared into the foliage. That frog knows how to jump and how to survive on land! How does it know?"

The wonder of such things has touched Sheldrake again and again since he was a child.

His topics are morphogenetic fields and the sense of the sacred. In his book *Natural Grace* he writes about how the separation of science, spirituality, and the sacred underlies our present crises of ecological devastation, despair, and disempowerment. Says Sheldrake in his book, "How else can hope in a new sense of meaning be awakened, if not by the coming together of those powerful traditions that were sent asunder in the seventeenth century? We need a new cosmology that speaks to our hearts as well as our minds."

We make ourselves comfortable in the library with cups of lemon-ginger tea. Sheldrake sits in a huge wing chair covered with a bright ethnic cloth, his long legs ending in feet tucked into slippers.

Years ago, I read Sheldrake's book *The Rebirth of Nature* and was fascinated by his approach. He shows us how we can think of nature as alive, rather than inanimate and mechanical and rational, which is how many mechanistic scientists have viewed it. His book explains how we have alienated ourselves from the magic, from the wonder in nature.

In his writing, I recognized that my own experience of reconnecting with that life and magic was being described by a scientist who received his education at several very renowned institutions, but who still managed to remain the wondering boy.

Sheldrake studied natural sciences at Cambridge and philosophy at Harvard. He took a Ph.D. in biochemistry at Cambridge in 1967 and was a fellow of Clare College, Cambridge, where he was director of studies in biochemistry and cell biology and where he conducted research on the development of plants and the aging of cells. From 1974 to 1978, he worked and lived in Hyderabad, India. After that, he lived for a year and a half at the ashram of Brother Bede Griffiths in South India, where he wrote his book *A New Science of Life*. Currently, he is a fellow at the Institute of Noetic Sciences in San Francisco. In 2003, he published *The Sense of Being Stared At and Other Aspects of the Extended Mind*.

Since reading his first work, I have wanted to meet the brain responsible for such thoughts. I am happy to settle down for a long talk with this outwardly peaceful man whose mind is so bright and inquisitive. What surprises me most about Sheldrake, I learn today, is both his acceptance of the world as it is and the balance he has found in it as he explores the connection between science and spirituality.

Our full day of conversation includes a long, guided walk on Hampstead Heath during which Sheldrake shows his eye for even the smallest detail in the world of nature. Upon my leaving, he encourages me to go on with my work, bringing people into deep contact with nature and their own inner being: "People have to experience their connection to nature. That is the only way."

What is nature?

Rupert Sheldrake: Nature is all there is. And I suppose nature is the entire universe, the entire physical world. In that sense, we are part of nature. You can use *nature* as an inclusive term to represent the whole of nature, which is the way it is used in natural sciences, the sciences of nature. That takes the entire universe as the subject of nature.

But *nature* is often used as a word in opposition to *culture* or to *humanity.* And then the word is used in a different sense, as a distinction between human beings and human culture and nature, which unfolds spontaneously, without human interference.

You have the same ambiguity with the word *animal.* Biologically speaking, we are animals. But people very often use the word *animal* to distinguish between humans and animals. So it is really a matter of usage of words. We have a lot of words that we can use in different senses.

Is it the usage of the word or the opposition or separation we have come to live with?

RS: I think all of nature is based on boundaries and separations. It is a feature of natural order. Every animal has a skin that separates it from the world outside. Every ecosystem has a kind of limit. The ocean has a limit in the shoreline. The forest has a limit in the tree line or the mountains. A cell has a membrane, which is its limit or boundary. So, if you look at the way nature is built up, it is built up of systems that have limits or boundaries, but which then are part of a larger system with its own limits or boundaries. For example, a cell has a limit or boundary as part of a tissue, which has a boundary of its own. An organ has a boundary: A kidney has a clear, distinct boundary. And then the whole body has a boundary. And social groups—like families—have boundaries, too. Even a flock of birds has a boundary.

We see the same thing in the way countries are organized. Parishes, counties, nations, and continents are different levels of boundaries and limits within that system. This is just the way that everything is organized.

When it comes to distinguishing between human beings and nature, then, of course we are part of Gaia. In another sense, there is a distinction between human culture in society and the natural world around us, which I think all human cultures have experienced—especially when people started the practice of agriculture. There you have a clear distinction between cultivated land, as in the fields, and uncultivated land, as in the forests.

So the way I like to talk about nature—that is, nature including humans—is perfectly natural to you?

RS: If we don't use those words, we need other ones to describe the distinction between the human and nonhuman worlds.

We seem not to know anymore that we are part of it all. We seem really separated—in our hearts, you could say.

RS: Well, in Western culture, there is that sense of separation, but I think that is only partial. One of the points that struck me years ago, and which I found reinforced ever since, is that most people in the Western world actually have two attitudes toward nature. One of them operates from Monday to Friday: the business attitude, which means to control and dominate nature's natural resources. But then the other one, which comes into play during weekends and holidays, is getting back to nature in a car—which is the ambition of the most affluent people in European society. What do people do who want to get rich by exploiting nature when they've made lots of money? They buy a country estate, so they can have a retreat in nature . . .

(Laughing somewhat ironically) I first noticed this when I had an acquaintance, here in England, who is a ruthless businessman. I don't think he cares too much about how he makes his money. With this money, he bought a country estate, which is a very typical thing in England. And on this country estate he has organic farming, he fights with his neighbors to preserve old oak trees, he's the defender of the natural world. He doesn't want people chopping down old trees on his estate. He doesn't want big roads coming through.

How do you explain this?

RS: Well, the distinction goes back into European history at least to the end of the eighteenth century, when there was a split in European culture. There was the movement of Rationalism, which has to do with science, the control of nature, domination through technology, and so on. And then, as a reaction to the alienating influence of Rationalism, there was Romanticism. The Romantics—the Romantic painters and the Romantic poets—were interested in reconnecting with nature. Here in

England, the poet Wordsworth is an obvious example. He lived in the Lake District, wandering around in wild places, and hated the idea of railways being built into the Lake District and millions of people pouring in and spoiling it all.

I think that the Romantic movement was a reaction against this dominating mentality. All of us are heirs to both these traditions. How it works in ordinary people's lives is that within the educational system and at work, rationalism prevails. It is the official philosophy of government departments, of bureaucracies, of universities, of schools and businesses. But most of us feel it as very unsatisfying.

Are we conscious of that other, Romantic part during the week? Do we really feel the frustration of having to wait for the weekend, or are we able to block out these feelings and live in a sort of schizophrenic way?

RS: I think Western culture is based on creating splits and separations. So, most people are quite happy working during the week to make money. But what people would think of as their "true self" is not a person in an office. At least in England, most people imagine that their true self is living in the country, cultivating hollyhocks in a cottage garden. This is what many people want to do when they retire. So they think that is their true self, even though it is a complete fantasy and totally unrelated to the way they actually live, which may be living in a suburb and working in an office.

Do you think they will ever reach the point where they can enjoy what they have been picturing for so many years?

RS: A lot of people do. They retire and move to the country, where they buy a cottage and grow hollyhocks.

If we go on in this split way, it's obvious that we'll see that it can't last forever, that at a certain moment the candy box will be empty.

RS: Oh, of course. But the reason that people don't feel this extreme alienation is that it doesn't happen all the time. Look at George Bush, for example, who wants to open up wildlife refuges in Alaska for oil drilling. And look at his poor environmental views in general. But I am quite sure that when he is on his ranch in Texas, he enjoys the broad skies and fresh air, and he probably really appreciates it.

Are you being ironic again?

RS: No! I think this is the normal, natural way that people in the West have come to live and have come to expect to live. I think it is an undesirable way of doing it, but I think it is the way in which the movers and shakers of our society—the people who shape our society and make the decisions—became rich and powerful. And those rich and powerful, at least in England, tend to go away for the weekend to the country and don't feel the disjunction from the country because they actually get quite a lot of unspoiled nature in their lives. They are able to believe that it is okay to have all this economic activity, because they can actually have both.

But, then, there is the boundary that you have mentioned . . .

RS: Of course, in overcrowded countries like England and the Netherlands, you can't possibly have everyone owning country estates and living like that. But this system works out also during holidays. People who make a lot of money want to go to unspoiled beaches in Sardinia or the Caribbean, or find an unspoiled part of Africa where they can go on a safari. Or they join an ecotour to the Antarctic, (laughing ironically) which is an unspoiled place until the ecotourist boats arrive.

Do you want to see this change at all? Or do you observe it and say, "Oh, well"?

RS: No, I think it is a bad but complex system. It's not as if one lot of people believe in exploiting nature and another lot believes in saving it. The situation is much more complex. They are usually the same people on different days of the week.

I certainly think it is a problem. But I don't think the problem is solved by convincing people that there is something beautiful and wonderful about the natural world. The problem is not so much to convert people to a new attitude as to find a way of breaking down the split that exists within people's own lives. In fact, an educated person is defined as being a kind of rationalist. Uneducated people are not rationalists, and that is why tabloid newspapers are full of things like miracles and psychic wonders and astonishing phenomena. And that's why serious newspapers like the *Guardian* and the *Independent*—and this is the same in all countries—avoid mentioning these topics, because they consider them to be superstitious and irrational. So ordinary people who have no intellectual pretensions can believe this stuff, but rationalists can't. And if they admit to belief in these things in public, people will classify them as superstitious, irrational, and so on. There is a huge taboo that has to do with the educational system and the social system. This taboo helps to maintain the split in educated people.

Did you suffer from that yourself?

RS: Of course. I was educated in an extremely rationalist way. I had a scientific education; I studied science at Cambridge. This is the pervasive atmosphere of the academic world. I spent fourteen years in the academic world of Cambridge and Harvard. Within that world, all of this seems normal. The reason I had to think about this continually is that a lot of my scientific work is an attempt to go beyond the mechanistic and materialist view and to expand the sphere of science in a way that reveals that, actually, we are more interconnected. For example, with the work that I've been doing on people's relationships with pet animals like dogs and cats, we have a situation where the split is particularly clear. Most pet owners keep animals because they actually want a connection with the animal, nonhuman world. There is no economic reason for people in Holland or England to keep dogs or cats. They don't usually need guard dogs. They don't need hunting dogs unless they are part of the very small minority that goes hunting. They don't need cats to kill mice in farmyards. And these people living in urban apartments with these animals keep them even though they are expensive. They are a nuisance, they are a lot of worry, and you have to pay the vet's bill. They keep them because there is some deep need to connect with the animal world.

How are you connected to nature?

RS: In the first place—in immediate family life, I mean—through my children and the animals and the garden. We have two sons, a cat, and a guinea pig. So in the house, there are all the normal things of teenage children.

I am pleasantly surprised that you see your children as part of nature!

RS: It is rather hard not to. (Jokingly) Everyone who has teenage children will agree!

I don't think that everybody sees them as part of nature. Most see teenagers as a problem!

RS: They are certainly that as well. But I think most people are aware that these problems are caused by hormonal changes, which are part of the growing-up process.

It may be that this particular form of rebellious behavior you get in teenage children in western Europe isn't even a necessary part of human culture. When I lived in India, I didn't notice this kind of teenage rebellion. It is a Western thing, having to do with prolonging adolescence. When I worked in Andhra Pradesh, in rural areas, the average age of marriage for girls was fourteen and for boys, seventeen. For a typical village girl, they organized a ceremony at her first period; they put up loudspeakers, there was a big party. She was sexually mature and within a few months she was a wife and within a year she was a mother. And in traditional societies, boys who pass the age of puberty have rites of passage and start to behave as men.

Here in the West, we've created this artificial extension of childhood where adolescents are not really treated as adults. They go on being students for seven, eight, nine years after reaching sexual maturity. This prolonged period is actually a very unnatural one in which people are not responsible and yet mature. In many ways, they want to be autonomous. So this rebellious behavior has to do with growing up and maturity and is basically biological.

In India, traditional farm children—little girls of eleven, twelve years old—are doing what most of them like doing. They are looking after babies, helping with cooking, gathering firewood, collecting water from the well, and doing womanly things. They are treated as responsible, useful members of the family, whereas children here are actually redundant: They are not economically productive till they are quite old, and they are dependent on their parents at an age where traditionally they would be contributing to instead of taking from the family. I think this frustrates the normal development process, which is to be a useful, contributing member of society.

After this discussion about growing up in Europe and elsewhere, Sheldrake continues to talk about the different ways he himself remains connected to nature.

RS: Second, there are the plants we are growing in the house and in the garden. There is the awareness of growing plants. Again, urban people everywhere have houseplants, even if they don't have a garden. I think, just as with the domestic animals, plants are a key part of this connection to the nonhuman world that people need.

And last but not least, we are exceptionally fortunate in living opposite Hampstead Heath, which is literally just across the road. In summer, I walk almost every day on the heath, where there are wildflowers, many trees, wild animals and insects of any kind. We have this large area of semi-wild nature right here.

As a boy, were you connected to that sort of nature?

RS: I grew up in a small town, where you could walk or cycle out of town and be in the fields.

When I was around seven, I could go out on my own. I used to wander off. Nowadays, children in England aren't allowed to go out on their own until they are much older. It is one of the great deprivations of the modern world.

Did you start your biology studies because of your love for nature?

RS: Yes. I was always specially interested in animals. I kept a lot of cats. My father was an herbalist and a pharmacist and he was an amateur naturalist, so I had a lot of encouragement from him. In the room next to my bedroom, he had a laboratory with microscopes. There he showed me all sorts of things: little animals in drops of pond water, the scales from butterflies' wings, hairs on leaves. He was a very good guide. He told me the names of the plants, the names of animals, and he showed me microscopic forms of life.

Was his introduction purely rational?

RS: It was rational, but it also involved a sense of wonder. His thing was more a cabinet of curiosities, like nineteenth-century naturalism. He had many cases of slides that I actually still have. They contain slides of bees' tongues, cross sections, stems of different plants. It was aesthetic more than rational. He wasn't writing academic papers or doing it for personal gain. He was an amateur in the real sense.

Could you relate that to the world you saw while on your bicycle?

RS: Yes, because I could relate the names of the animals and plants to that. And every year, I used to collect tadpoles and I would grow them into frogs, (laughing) as I still do! I also used to collect caterpillars and feed them and see them turn into butterflies. I was always interested in transformations.

When I look at plants, what I have studied and how I look at them is: How do they develop their form? How does the leaf form? How and why do the veins of the leaf form where they do? How do the cross veins join up? When I look at plants, there is the constant wonder at how these amazing forms develop. [They make] me think that the world is full of beauty, wonder, and diversity. I used to work at the leading edge of developmental biology in relation to plants but grew disillusioned. The conventional approach leaves out almost everything that really makes plants what they are: their individual forms and shapes.

Can you say that plants have pushed you to research more and more, because you knew from them that so much was possible?

RS: I think they made me realize something about the way nature is organized. They made me realize that science as we have it, mechanistic science, is radically incomplete. It is dealing with only a small proportion of what is there in the natural world. Plants strengthened the sense that there is a huge amount of things we don't understand, even of ordinary, everyday phenomena; that science is really at the beginning, not near the end, of an understanding of nature.

Yesterday, I saw that one of the small toads had just completed its transformation. It had come out of the water, ready to go. So I took it down to let it go in the garden. I held it over some wet grass, and this little toad just bounced off in one great leap and hopped off into the grass. It had never jumped before, and it had no one to show it what to do. It had to rely on instincts alone, in a totally strange world. And yet a lot of them survive in a world of uncertainty!

The admiration that shines through his description suggests this was the first time ever that Sheldrake witnessed such a natural transformation and illustration of animal instinct. In reality, this is what he has been doing ever since he was a small, wondering boy who wanted to know everything about the natural world. His enthusiasm is more than contagious as he continues his story of wonder and admiration.

RS: So, the spontaneous, instinctive behavior of animals is what is astonishing to me.

Does this side of the animal world give you the strength to carry on, or at least feed your romantic side?

RS: I don't think it is particularly romantic. It's just the way nature is, the way the world is. As a scientist, I am interested in how we can try to understand this. But as an ordinary person, I am just interested in seeing all these things. I found it inspiring through the sense of the spontaneous order and inherited wisdom that seem to be built in so many animals.

The way in which you're observing must affect your personal life, too. Those animals are probably teaching you a great deal.

RS: It does have an effect on what I do. For example, I am a vegetarian, although we do eat fish. But we don't eat mammals or birds, partly because I strongly disagree with factory farming. I have very strong objections to the modern agricultural industry. And most meat that you buy in the shops and supermarkets is from animals grown in totally unnatural conditions. If I did eat meat, I would eat game and animals that have grown wild and free. I would probably find it much easier to eat that kind of meat. So we eat organic food, drink organic milk, and don't eat meat. Nothing special about this, because millions of other people are doing the same—at least in Britain.

(Suddenly, with some concern) I don't know if you eat fish. (After we respond affirmatively) Oh good! that's what we're having for lunch!

Lunch is served in the lovely garden surrounded by huge trees at the back of the house. Both Sheldrake and his wife, Jill Purce, work from home as much as they can.

With our stomachs full, we borrow his sons' sneakers to wear on our walk over Hampstead Heath. We listen to an animated Sheldrake as he shows us a four-hundred-year-old oak and talks about all the people who love to be out in nature, to be in the sun, during holidays. He comments that actually what they are unconsciously performing is a sun ritual. They are adoring the sun. If only they would just consciously connect!

Back at the house, I bring up the subject of Sheldrake's theory of morphogenetic fields. This is from his book The Rebirth of Nature:

> Morphogenetic fields contain a kind of collective memory on which each member of the species draws and to which it in turn contributes. The formative activity of the fields is not determined by timeless mathematical laws—although the fields can to some extent be modelled mathematically—but by the actual forms taken up by previous members of the species. The more often a pattern of development is repeated, the more probable it is that it will be followed again. The fields are the means by which the habits of the species are built up, maintained, and inherited.

How do morphogenetic fields influence your personal life?

RS: The theory enables me, when I look at animals and plants, to see in them a kind of memory of what has come before, the presence of the ancestors working through them. So it affects the way I look at plants and animals and people. It affects the way I look at and experience social groups, such as families, because when you see your family as a field, you see how the different members are interacting through a field. Also, there are archetypes. Those fields have patterns from the past, a kind of memory pattern. I feel them at work in my own life and in the lives of those around me.

Many of the things we are talking about seem to me related to "oneness." Do you agree?

RS: I would say diversity within oneness, because with the whole nature of nature as a holistic worldview, you've always got unities that contain diversity. Those unities themselves are part of a larger diversity. Gaia, for example, stands for the unity of all life and all things on Earth. But the Earth is part of the solar system; and the solar system is part of the galaxy. At each of these levels, there is a kind of oneness. The whole solar system is interrelated through the gravitational field to the sun. And the sun is part of the entire energy-flow system and organizational structure of the galaxy. There are many levels of oneness.

Do you see that anything anyone does has a resonance in that huge oneness?

RS: It must be so, at some level. The only question is: To what degree is it significant or important? Newtonian physics states that all particles attract all other particles in the universe. If I move a cup from one place to another (he moves his cup), this is now slightly affecting the entire universe! But in this case, the effect is negligible and completely trivial. The key question is: Which actions have a significant effect on others?

Can thoughts have an effect?

RS: Thoughts too, I think, will affect what goes on around us, and potentially affect other people—but again, most of them perhaps in a small way.

Are you familiar with the work of Masaru Emoto, who shows the enormous power of our thoughts and emphasizes that they can transform something sullied into something beautiful?

RS: I'm afraid when it comes to Masaru's work, I revert back to my more skeptical, scientific mode. I have met Masaru and I've seen his work—but I question the technical side of it. I'm sorry to be boring about it, but we know that every water crystal is different—Masaru himself says so in the introduction to one of his books. If you take a sample of water and you crystallize it in a dish and deep-freeze it, of course crystals are going to be formed. Masaru photographs only one of those crystals without a statistical analysis, so it may not be representative. That is why I see his work more as a form of art than of science.

He came here and recorded my wife doing overtone chanting—she teaches a form of Mongolian overtone chanting; it's a form of meditation. Masaru played this music to water and showed a picture of a seven-sided crystal, an unusual form of water crystal. But I am sure they weren't all seven-sided; I mean, I suppose out of hundreds he picked the one that showed an unusual form.

Many people have found his work helpful as a kind of inspiring view that thoughts can affect water. And maybe they can. I'm not saying Masaru is wrong; I'm just saying that his scientific evidence is not terribly persuasive. If what he's saying is basically true, he could make a stronger case if he included some statistical analysis. And if what he's saying is a kind of artistic, subjective, romantic view of water and how we can affect it, then still this is perhaps an important message, but it's not the same.

Sheldrake himself does research on disputed topics, that skeptics attack, and knows that unless he can show the evidence to be solid—that it stands up to skeptical scrutiny—it will be dismissed.

He tells us about organizations whose job it is to maintain the rationalist worldview. They are self-appointed vigilantes who "patrol the frontiers of science," as Sheldrake puts it. He is convinced that if Masaru Emoto's work were subjected to the scrutiny that his undergoes almost daily, Masaru would be torn to shreds.

RS: Probably, they have done so already, in their skeptical journals, which means that Masaru's work will stay in the New Age enclaves; it can't possibly become mainstream unless it can stand up to that kind of attack. That's a shame, because I think the general principle of water being influenced by thoughts and intentions may well be a valid one. And the traditions around the world of holy springs, holy wells, and holy water are deeply rooted. I think the reason his work sells in all parts of the world is precisely because there are already traditional beliefs that water can have qualities that modern science ignores.

As a healer, I see how focused attention can influence matter. I am curious whether you have done any experiments with this.

RS: Indirectly, I've done research—for example, somebody thinks about calling somebody and then that other person phones, or a dog knows when his owner has decided to come home and goes and waits by the door. There you have a distant effect that is measurable. In a sense, a lot of my work has to do with that. Research has been done recently on healing through prayer. This includes double-blind, controlled trials of heart patients and AIDS patients in a traditional scientific setting. They are prayed for by people who get their photos and names. There is significant improvement in the health of the prayed-for compared with those who haven't been prayed for. So I think that focused attention can indeed have an affect.

My feeling is that it would be best if, instead of thinking, "The Earth is so polluted," we would think, "It's a wonderful world." That would make a difference.

RS: I agree. I believe that attitude makes a huge difference. But there are personal differences, too. There are some people who are always gloomy, aren't there? In Winnie-the-Pooh, Eeyore the donkey is always miserable and gloomy. There are people like that!

Could you say we're responsible for the state of the world around us by our way of thinking?

RS: We are, up to a certain point. But within a society or group, there is always going to be a balance. If there are too many optimistic people, there will appear those people who look at everything in a gloomy way. You have to look for the right balance.

Because we are talking to a man who is close to animals out of a genuine interest in them, not merely because of his profession, I'd like to know more about how he sees our relationship to animals.

What can we learn from animals? Do you think they are telling us something?

RS: I don't think they are telling us one simple message. I've also learned about animals from what other people have told me, especially when I was writing one of my books [*Dogs That Know When Their Owners Are Coming Home, and Other Unexplained Powers of Animals*].

I feel that animals are interconnected with us and with each other telepathically and have connections that work at a distance, which has a biological function. Because animals live in social groups, they need to be coordinated with other members of the group. That is the one point. The second aspect of my book deals with their sense of direction: the way animals are linked to places—the way that pigeons, for example, can fly to their home from hundreds of miles away and migratory animals, such as birds, butterflies, and turtles, can travel thousands of miles with this astonishing sense of direction. Animals have particular relationships with particular locations. There is a locality in the animals' existence that we don't yet understand. I think this involves morphic fields that connect them to their homes. Take animals away from their homes, and the field that connects them is stretched . . . Mobile phones will help us to develop our telepathy further.

So what are animals teaching us?

RS: That there is a connection to place that is an important part of animal nature. Human beings have it, too: We have a sense of "going home." One reason we have so many American tourists in Europe is because many of them, even several generations later, want to come back to their Old Country to see where their roots are. We see this through the whole of human history—people have [always] gone to special places for pilgrimages. There is a connection with sacred places, places with power. All traditional cultures have pilgrimages, where people feel that certain places are particularly important: a sacred mountain, a holy well, or a shrine where a great man or woman died. The Muslims have Mecca, which is celestial stone—the Kaaba, a meteorite—to which they all turn to pray. Here in England, before the Reformation, we had many pilgrimage places. The Reformation abolished pilgrimage and sacred places and I think it had a traumatic effect on England and everywhere else it occurred. Surprisingly, it didn't take long before the same instinct came out again, this time disguised as "tourism." Pilgrimage has been secularized into tourism, which again is based on the sense of connection with place

Where do tourists go when they come to London? Often, they go to Westminster Abbey, Saint Paul's Cathedral, to all the ancient pilgrimage places. In Paris, they go to Notre Dame; in Japan they go to the Sacred Mountain. But people can't relate to the place in the direct spiritual way; they have to relate to it through facts and statistics: This was built in so-and-so, contains so many tons of stone, and so forth.

The third thing I deal with in this book is premonitions: For instance, some people know when earthquakes are about to happen. It is a sense of connection to the environment—not just to a place, but to events that are going to happen. The fact is, there is something in our animal nature that enables us to anticipate disasters or catastrophes.

This animal nature forces us really listen to our intuition, which asks us to open up to the Earth, to feel the Earth.

RS: Yes. In fact, most people have this already. But it's usually repressed or suppressed by a kind of rationalist view.

Or by their busy lives.

RS: That is another reason that people are not recognizing these connections.

As strange as it may sound, hunters are traditionally very close to nature, because they have to know the habits and instincts of the animals they are hunting. It makes them much more connected to nature.

How do you consider the question of awareness, consciousness, or even soul in all that lives? I know you have talked about the awareness of the sun before. Could you tell us a little bit more about that, for a start?

RS: One of the things I'm interested in is the possibility that there might be consciousness, or mind, associated with heavenly bodies.

If we think of Gaia as alive, that still leaves the question of whether she has consciousness. It is possible to be alive without thinking. After all, plants are not thinking, but they are definitely alive.

But then there is the question of whether the sun is conscious. All traditional religions have seen the sun as a god; the sun is a divine being in those religions. And when children draw a sun, they draw it with a face and a smile. So, the idea of the sun being alive is very ancient, found all over the world.

It may be impossible to refute the idea that the sun is conscious. But probably it will be as impossible to prove it! There is still much we don't know. Why do masses of people go and lie in the sun on holidays? It's the old sun worship, really, if they would only be conscious of it!

Are there more scientists open to this possibility or are you alone in this?

RS: Together with some friends, we organized a small symposium a few years ago under the title "Is the Sun Conscious?" For everyone present—people representing different sciences—this was new territory. We asked things such as "If the sun thinks, what does it think about?" And then we arrived at the conclusion that the sun is the heart of the solar system, so maybe one of the sun's main concerns is the state of that solar system. Or the sun might even communicate with other stars in our Galaxy, and the Galaxy itself might have a kind of mind. Of course, all of this is speculative.

Would you say that all life carries a consciousness of itself?

RS: It is a problem with words, you see. I think the word I prefer is *soul.* In ancient Greece and the Middle Ages, Aristotle and then Saint Thomas Aquinas had the idea that all nature was alive. And what makes a thing alive is a soul. Plants have a soul that gives them their shape, animals have a soul that gives them instinct, and the human soul includes, in addition, a rational or intellectual aspect.

Descartes said the whole universe is a machine with an external designer. In his system, the only soul left in the whole natural world was the rational mind of human beings.

At the end of one of my most extended conversations—we started early in the morning, and now that we have come to the last part of our conversation, this splendid day has almost gone by—I am excited to ask Sheldrake our last question:

What is love?

RS: There are several different kinds of love. It is not a single thing. It has to do with the bonds and connections between organisms, including parental love, erotic love, and love between members of wider social groups.

I think the most basic forms of love are the bonds and the protective behavior between parents and their offspring. For instance, birds will incubate their eggs, collect food, feed their young, and protect them when they are attacked. Sometimes they attack the predators, even if they can be killed while protecting their young. Now, in that you have all the behavior associated with love: parental care and altruism.

But some people will say that this doesn't count as love, because only people can love and love has to be conscious. Again, it is a matter of words. If you want to define *love* as an exclusively human phenomenon, you rule out all this altruistic behavior by animals, and mammals and birds, in particular, care for their offspring. Reptiles, on the other hand, don't have a social bond; they are usually solitary. Turtles, for example, leave their eggs behind, and once the small turtles come out of the eggs, they are completely on their own and have to find their way to the sea without the guidance of their mother.

Love has to do with social bonds. The most fundamental kind is parent-child love. Interestingly, in many religions it is this parent-child metaphor that is the model of divine love. In the Christian tradition, it is the father and the son; in the Catholic orthodoxy it is the mother and the child. So also within religion you see this idea of parental love as the most fundamental kind.

Works by Rupert Sheldrake

Chaos, Creativity, and Cosmic Consciousness (Rupert Sheldrake, Ralph Abraham, and Terence McKenna). Rochester, Vt.: Park Street Press, 2001.

Dogs That Know When Their Owners Are Coming Home. New York: Crown, 1999.

Natural Grace (Rupert Sheldrake and Matthew Fox). New York: Doubleday, 1996.

A New Science of Life. Rochester, Vt.: Park Street Press, 1995.

The Physics of Angels. San Francisco: HarperSanFrancisco, 1996.

The Presence of the Past: Morphic Resonance of the Habits of Nature. Rochester, Vt.: Park Street Press, 1995.

The Rebirth of Nature. Rochester, Vt.: Park Street Press, 1990.

The Sense of Being Stared At and Other Aspects of the Extended Mind. New York: Crown, 2003.

Seven Experiments That Could Change the World. Rochester, Vt.: Park Street Press, 2002.

Web site: www.sheldrake.org

JANE GOODALL

I prefer to use the term
sustainable lifestyles.
That is all we need to aim for.

Jane does not have time for an interview, at first. Her schedule leaves her only a few days to spend with her grandchildren in the United Kingdom, and that, of course, is sacred. But then one of her editors wants to pay her a visit, and she decides to combine this visit with my interview, giving us an hour or two for our conversation.

We find her not in her own home, but in the flat of her secretary in London. It does not seem to matter. The material side of life does not interest her, anyway. She travels around the world in simple clothes and only the simplest makeup. A badge for Roots & Shoots, her service learning program for youth, is pinned to her dark blue collar. Her gray hair is caught in a ponytail. She is pale, thin, and a little distant—very much the observer—and she exudes a quiet peace in spite of the traveling, lectures, meetings, correspondence, and demands of a global nonprofit organization, the Jane Goodall Institute, based in Silver Spring, Maryland, which, besides Roots & Shoots, includes the Gombe Stream Research Center in Gombe National Park, Tanzania.

Nature is her real home. She feels more connected to spiritual forces when in the wilderness. It takes her back to her childhood, when her dream of Africa was inspired by *Doctor Doolittle* and pictures of Tarzan that she saw when she was eight years old.

A rich inner life guides her. Her issues of concern are multilateral—poverty, youth, animal, the environment, and politics—and her approach is holistic. Many people are familiar with her work with chimpanzees, our closest genetic relatives, through documentaries and movies made about her life and research. It was Louis Leakey who first sent her to study these primates. "They are like us, superb ambassadors who help us understand that we are part of nature," she explains. They function as a mirror for humans. Jane's research has bridged the link between our two species, making clear that we humans are also animals, part of the web of life. During our discussion, I offer that animals behave far more in tune with their environment than do humans: They don't kill merely for the sake of killing; they don't destroy for the sake of destroying. Jane immediately puts a halt to my remarks, however, pointing out that chimps do consciously kill and destroy, and that this shows us even more how closely we are related.

Despite thousands of interviews and talks she has given in these past decades, we embark on a lively conversation about the essence of nature.

What is nature?

Jane Goodall: What is nature, really? (Laughing) Well, I would say everything that is not man-made is nature, including ourselves. People think of nature in different ways. But . . . nature and the natural world are different, because the natural world is a place completely unspoiled, like an unspoiled ecosystem. And nature would even be a little wheat that grows up in the middle of the inner city. That is the way I would see it, anyway.

So, you make the distinction between the natural world and nature—a distinction in the sense that the natural world is part of nature, but not the other way around. The natural world is something special.

JG: Yes, the natural world is where we haven't interfered with a whole system. Or even if we interfered once, we have stopped interfering and left it to itself.

Where is this type of unspoiled nature still to be found?

JG: Oh! I have just come from the most amazing place in the heart of the Congo basin. It is a twenty-five-kilometer [15.5-mile] walk to even get there! It is a forest area called the Goualougo Triangle, in the Nouabalé-Ndoki National Park, where even the Pygmies have never been. It is now protected.

Protected by whom?

JG: It is a national park protected by the Congolese government. And there are other forests like that.

What is your connection to nature?

JG: In nature, personally, is where I have always felt at home. It felt like that whenever I was out in the woods or the forests or the Serengeti Plain and away from people—or at least away from most people—but in particular away from the noise, the man-made noise . . . away from the ugliness that we often surround ourselves with, and the roads that cover more and more of the land.

My connection with it is that when I am in it, when I am in a place like that, I feel much more connected to the great spiritual power. It seems that that great spiritual power is more accessible to me when I am away from cities.

Does it inspire you? And if so, how?

JG: Oh, yes, it does. I think it takes me back to that stage of childhood when everything is wonderful, everything fills you with "Ooooh!" I have never lost that.

Where was that childhood?

JG: That was in England. Our garden was wild and big and the cliffs along the seashore were untouched. Of course, since the war, it has become more developed. At that time, it was very wild. I used to go riding out in the country. Again, it was farmland, but it was still a vast nature. So I think I never lost that.

At that time did you already want to escape from people and their noise?

JG: Yes, even as child of about eight years old I was writing poems about being away and being out in the wild.

Eight years old is very young. Why do you think it started at such an early age?

JG: It started long before! It started when I was one and a half years old, when I watched worms crawling around my bed. Mother says that maybe I wondered how they crawled without legs. I took them to bed. We lived in London, and my mother—instead of saying "Ugh, get those dirty things out of your bed," she said, "Jane, if you leave them here, they will die because they need the earth."

So your mother encouraged you to be with nature?

JG: She was amazing. And there was another time, when we went to Cornwall. I was about three, I think, but I remember it quite well. We went to a rocky beach that had so many little living shellfish. There were some beautiful yellow ones, and I was totally entranced. They were so beautiful! I was collecting them, and nobody realized that I had a bucket full of living little yellow sea snails. I took them back to my room. I must have known that I shouldn't have done it, (laughing) but I didn't know that they would die! I thought that I could have them in my room. And Mom came in, and everywhere there were snails crawling, all over the walls and on the bed.

Somebody said to me: "They are all going to die without the sea." Apparently, I became totally hysterical, and the whole family had to come help catch every single one and we took them back to the sea. So, it started really early. And then, of course, the famous story of when I was four years old and I hid in a henhouse for hours to see where an egg comes out. As though it was yesterday I can still remember watching the chicken lay the egg.

What was it in you, do you think? Was it curiosity, or wanting to belong to that world?

JG: It was a mixture. I wanted to know more about it. I don't know why I had this desire, or where it came from. It seemed to have been born in me.

Was it more than just curiosity?

JG: A lot was curiosity, but wonder as well. I know that my nanny—I must have been about six or seven—saved up coupons from cereal packages. And in those days, when it said "free," it was really free—it wasn't that you had to collect and send a postal order for twenty quid with it! The coupons got us a book—I still have it. It was called *The Miracle of Life*—not meant for children. It was a complete, scientific description of most everything. It had chapters like "Many Tongues for Many Purposes" and "Many Feet for Different Purposes." I guess it was like a school textbook for high school students. It was big and fat and it had black-and-white illustrations, photos, very detailed. It was the history of medicine, the history of discovering anesthetics, as well as nature. And that was my bible, really. That and *Tarzan*. (Laughing) I was very jealous of Tarzan's Jane!

Is that why Jane became your name, or was it already Jane?

JG: It was Valerie Jane. It was changed not because of Tarzan's Jane. I hated her! I would never change my name to hers.

So it was really Tarzan who brought you to Africa?

JG: Yes, he and *Doctor Doolittle*. I actually have the book with me. I was reading it to my grandchildren yesterday.

She leaves to look through her bag in another room and returns to our table with the book in her hands.

JG: I got this book when I was eight, you see. It was the first book I ever owned, given to me by my grandmother. She paid six shillings, secondhand. It was printed in 1942. This picture has stayed with me, and that is really what took me to Africa.

(Opening the book) This was the picture. (She reads) "The bridge of monkeys, which had never been seen by a white man before, over which the doctor—" I mean, how improbable, that you can walk over monkeys! But it didn't seem so then, when I was eight . . . "So with his little medicine box, his pig and the other animals, they escaped from the wicked king—" And then the picture says: "John Doolittle was the last to cross."

And then of course, *The Jungle Book* did the same thing. It was about being in the wild, being with animals, and being part of nature, part of the natural world. That is what I wanted to do; I wanted to live with animals and write books about them.

How old were you when you knew that?

I was ten at the time.

And how old were you when you went to Africa?

JG: I went for the first time when I was twenty-three, when a school friend invited me. That was the opportunity I had been waiting for.

There was still no money for this. I had a job in London and I had no degree because we couldn't afford university, and you couldn't get scholarships in those days unless you were good in a foreign language, and I have always been hopeless. So, my mother advised me to do a secretarial [course], to be able to get a job anywhere in the world. I did that and had a job at a documentary films company in London. The same day that my friend's letter arrived, inviting me for a holiday, I gave my notice. But there was still no money, so I worked as a waitress in Bournemouth, probably some five months, before I left for Africa.

While I was in Africa, I met Louis Leakey, and that was it. That was the sequence.

And you stayed?

JG: No, I had to go back to England, because it took Louis over a year to find money for me to study the chimps. But I stayed in Africa that first time for a year and worked for Louis at the museum as his assistant. I was able to go out into the Serengeti when it was totally, completely untouched. It was just wild. Where we went, there wasn't even a track from the previous year.

Every evening after digging for fossils, Juliana, another English girl, and I were allowed out onto the plains. We saw so many animals. And we walked among them! Amazing.

Your dream came true.

JG: Yes! And I felt at home. That's the funny thing.

Richard Leakey—who was quite opinionated even though he was a boy of fifteen—he and his big brother came. They had been going on digs, they had been brought up their whole lives in this environment. There I was, straight from England, really. And yet even Richard seemed to follow my lead or ask what I thought we should do, like when we met up with a rhino. It was really strange, looking back, that they seemed to sense that I really was at home.

And you knew how to be with the animals?

JG: At that time, I had no fear at all. Fear came only when I had a son. Very funny, but subconsciously I did change after becoming a mom. I became a totally different person; I couldn't believe it. All of a sudden, I was a protective person.

The first time I realized that I had changed was when I was in Ngorongoro crater, studying hyenas with Hugo [Hugo van Lawick, her husband at the time].

One day, my little baby was asleep in the back of the car and an elephant walked across the crater. Before Grub was born, I would have been so excited that the elephant was coming: How close will he come? But now, all I could think was "What is going to happen if he comes too close?" A complete change. It must have been motherhood.

It is part of nature. You see it with the animals too.

JG: It is! Especially with mother animals.

Was it a temporary change—only when your son was little?

JG: No, I think I lost a certain naïveté or innocence. It is a matter of having new responsibilities, because your children will always be your children, no matter what age. You never get back to that complete freedom, when it is only your life and you are selfish enough not to be bothered about your mother, who is feeling exactly the same as you will later. But you don't realize that, then.

Did you gain something too, because you could probably understand the animals even better than before?

JG: Oh yes! I understood the chimp mother infinitely better, absolutely, when I had my own baby. It made a huge difference—understanding funny things in the chimp's behavior that I had not been able to understand [before].

What made you choose the chimps, or did they choose you?

JG: Louis Leakey chose them for me, because the chimps, being more like us than any other living creature, really are the most superb ambassadors from the animal kingdom to help people understand that we are indeed part of nature. There isn't a sharp division between humans and animals. I never saw it, but, of course, Western science has traditionally perceived this huge gap—as has Western religion. This misunderstanding is missing from Buddhism, Hinduism, and the Native Americans. It is particularly Western. I think the chimps, more than anything, have forced many scientists to reevaluate their belief systems. And then, once you have a new respect for the chimps, you realize: Well, differences between them and us are not differences of kind, but of degree. And once that line becomes blurred and there isn't an impossibly unbridgeable gap, that leads you to a new respect for all the other animals.

What degree of difference?

JG: Well, for example, language. We are the only creatures with a sophisticated spoken language. But chimpanzees and also other animals, especially the other apes, show many cognitive abilities we used to think unique to us, although we have developed them to a higher level—such as symbolism, understanding abstract symbols; being able to use them in communication; having a "theory of mind," as it is called, when you understand the wants and needs of others.

Are we the only species that thinks about thinking?

JG: Yes, I would suspect we are. It is a difference of degree, not of kind. They [animals] think. We used to believe that they didn't, though I never thought so. I always believed they did think, because all through my childhood, I had a wonderful teacher: That was my dog, Rusty . . . He was a very unusual dog. He was meant to be with me. He was an extraordinary dog. In the long string of subsequent dogs I've had, there has never, ever been one that even approaches Rusty.

What did he teach you?

JG: He taught me that animals have very, very vivid personalities, very strong emotions, a sense of right and wrong, and the ability to reason and work things out.

Can you tell us more about this?

JG: Chimps certainly seem to have that "lightbulb" experience that tells them: Ah! That is the solution. They try different things to solve a problem, and then suddenly, they find it. A perfect example is a captive chimp in the colony of this amazing man called Wolfgang Kohler, an Austrian. He had a colony of chimps on the Canary island of Tenerife. His work was largely ignored. He knew so much about his captive chimps and people disregarded that. They considered them contaminated.

His brightest chimp was called Sulton. The problem was a simple one: There was food out of reach. And the solution to get it was to use a stick. Sulton was hungry and he wanted that food. He tried but couldn't reach it. Then he looked around and saw a little twig—he clearly got the idea. He reached out with that, but still he could not get to the food. And suddenly, he ran all the way back to the spot where he had been brought in—where they had deliberately laid a long stick that morning. You could see him thinking: I need a stick. Ah! I saw a stick. And off he went to find it. He remembered!

Kohler said that from the moment the solution becomes apparent, the whole behavior changes. The chimp stops looking around for something to solve the problem with; he just goes ahead and solves it—all in one fluid movement. Kohler says this chimp was clearly thinking. And Rusty thought too.

Do they also think into the future and into the past?

JG: Chimps can think into the future and certainly into the past.

But don't they think about thinking then? Is this the difference between their thought and human thought?

JG: I don't think memory is thinking about thinking. Memory is being able to draw on something in your mind that gives you a solution. That is, what worked yesterday is going to work again. Take the chimp again: Some chimps were in a zoo, in that famous Arnhem colony in Holland. They were shut outside during the day, and one autumn day, it suddenly turned cold. One chimp was old and didn't have much hair, so she was shivering. And then the staff let the chimpanzees into their cage inside. The next morning, the chimps got their breakfast, and before the doors opened and before she could have any knowledge of the outside temperature because it was warm where she slept, this chimp went over to the door and then suddenly left the others and turned back. She got a huge armful of straw that she took with her when the door opened to let them out.

So this one example shows that animals are able to look into the past and into the future.

JG: Yes, but it is an anecdote and not a scientific experiment, so some would say we shouldn't pay any attention to it. In fact, especially with creatures as individually different and amazing as chimps, if you collect the anecdotes, you begin to get a sense of the whole, rich flexibility of chimpanzee behavior and capability. Maybe that particular behavior occurred in our human background, too. Maybe it gives us an idea of perhaps how early men moved in the direction they did. I think that anecdotes are very, very important.

You were not a scientist when you started your work with the chimps?

JG: No!

You had to become one?

JG: Yes. Louis Leakey got me the money for a year and a half. And he said, after I had been out there for about a year, "Jane, you have to get a degree, because I shan't always be around to get money for you. You have to stand on your own two feet. We don't have time to mess about with a B.A., so you have to go straight for a Ph.D." I went to Cambridge.

You had to leave your beloved continent and go back to the United Kingdom again?

JG: The leaving and coming back wasn't the problem. The problem was that when I came to Cambridge, I was greeted with a real hostility. I found very few people whom I could talk to.

Why, exactly?

JG: Because I had named the chimps instead of numbering them. I talked about them having personality, which most people thought was unique to us. I talked about them having emotions and I talked about them being able to think, having a mind, being intelligent. And all of this was not scientifically acceptable.

Instead of them reacting with "How interesting!"

JG: They thought the way to learn about animals was by blinding them, deafening them, and doing horrible things to them to see what would happen if you destroyed a piece of a system. In other words: You take a clock slowly to pieces to find what makes it tick instead of waiting and observing the clock and examining the inside without touching it.

For my unwinding, there were two people I would spend time with to get away from the animal behavior unit at Cambridge. I used to talk to Desmond Morris, who later wrote *The Naked Ape*. I knew him when he was the curator of the London Zoo, because I worked there the year before. And I would talk to Roger Short, who was at the vet school. They were doing a study of wild deer in Scotland. They really were people whom I could relate to, who were open to what I had to say about animals.

Still, the first scientists who were really, really interested in everything I had to say were the child psychologists and psychiatrists.

Do you think that is because they are open to children, to allowing the child to speak or behave how it will, or . . .?

JG: No, I think it is because they weren't totally brainwashed by the very reductionist thinking of the ethologists of that time—not the founders of ethology, like Konrad Lorenz and Nikolaas Tinbergen, who were open to wonder. I mean the ethologists who tried to make ethology into a science, a coldhearted science, which got colder and colder. You were required to reduce everything to its lowest possible cause.

What did that do to you?

JG: Luckily, it didn't do anything to me. By that time, I was twenty-seven, nearly thirty. And I had Rusty!

I imagine it could have broken something in you . . .

JG: But it did not. What it did do: It taught me to think logically. I actually loved it. I had a wonderful supervisor. And I'll never forget that I had written to him about the chimp named Fifi and her newborn brother, Flint. Every time anybody came to try and look at her little brother, Fifi would go (imitating chimpanzee vocalization while hitting a fist against her chest), and she would threaten them away. So I said that Fifi was jealous. And Robert Hint told me, "You can't say that." So I asked, "What shall I say then?" because she was jealous. He told me to write this: "Fifi behaved in such a way that, had she been human, we would say she was jealous." It was just playing with words.

So, I was very fortunate in who and what I had around me. I had my mother, to start with; then there was my grandmother, too, and the place where I grew up, which was relatively wild. And the tree: I took all my homework and all my problems up into the top of the tree in the garden. And I had my dog, Rusty. Then I had Robert [Hint] at the university. And, of course, there was Louis Leakey. And I had Tarzan and Doctor Doolittle.

Good company! So, do you think it is at all possible to have contact with nature through science?

JG: Yes, absolutely. I don't think that science in its essence needs to be cold, hard, and reductionist. Science, in its true form, is wanting to learn about, wanting to know about. With that attitude, you will approach nature with a curious mind and a sense of wonder—people like Einstein did not lose that! And this way, of course, I think your appreciation of nature can even be heightened, because the more we learn about nature, the more truly wondrous it turns out to be. I mean, my sense of awe becomes deeper and deeper the more I learn about it.

Are you still working scientifically?

JG: No, right now I am not. Not really, although when I get back into scientific conferences and I hear my students who are using my old data, I sometimes have those longings.

Did you come to understand the chimps more by using a scientific approach or by your own experience?

JG: By my own intuition—observing, watching, waiting, and letting my mind play with it until suddenly I get that feeling of "Oh, that is what that must mean." Or I get that feeling because I see the same thing eight times in the same context. So you realize what that sound means, and why she does this or he does that. And then, because chimps are continually demonstrating completely new behavior, you realize how flexible they are and how much an individual can contribute to the history of the group, and how mothers pass down traditions so they become part of culture. A lot has to do with the curiosity of the scientist, which involves a lot of self-discipline. You don't just jump to a conclusion.

You may, actually, but you must then test it. And that is what I find fascinating.

Do you think you would have reached the world in the way you have if you had not done scientific research?

JG: No, I'm pretty sure.

What is the balance you maintain between science and your own intuition? Is there a balance or are the two completely integrated?

JG: I think they are totally integrated. I think they always have been, actually—that is, as soon as I got to understand that science matters and got over my first frustrations. Everything in my life, I feel, has been laid out for me. All I had to do was make the right choices.

You are a busy woman. How do you find balance between the quiet and peace you need and your very busy life?

JG: I think it is because I can easily find peace inside. After being in the forest for so many years, it has become a part of me. It is something people are continuously saying to me: How can you sit there and seem so peaceful? People ask if I meditate. Officially I don't, but I think I do in my own way. I know the forest is there, and now I know that the heart of the Congo is there too, because I have experienced it. It stays within. I got back just two weeks ago. All the rest of the team got horrible infections from the bugs, they got malaria and everything, and I was fortunate that I got nothing!

Did you feel you were home again there, as you had the first time you went to visit Africa?

JG: Not during the twenty-five-kilometer march; that was an endurance test. It was horrendous, actually, and I was told it was twelve kilometers [2.6 miles]. Mike Fay is the one who walked two thousand miles across the Congo basin, and he really wanted me to go and see it for myself. He said, "You will be more passionate if you see all of the forest." I said, "Mike, that can't be. I am passionate."

But I think he was right. (With love, respect, and worry in her voice) It was not just the forest, but all that he was able to tell me about the dangers that these forests are in. So the combination of the two makes me really glad that I went. And the feeling of being in the middle untouched nature was magic. It started a lot of things inside me.

We are all silent for a moment before we continue.

Do you still go to Africa often?

JG: No, I don't anymore. I go only twice a year.

Does that mean there is no longer a need for you to be there in order to be able to do the work you do in the rest of the world?

JG: No. There is so much work to do in the West and the East. Don't forget places like China, Japan, Hong Kong, and now India. So, there is no time. I'm lucky that I don't have to go back to recharge too often. Then my life would be even more hectic. This trip I just had in the Congo is the longest I've been away from civilization in years.

Do you see all of this as a task?

JG: A mission, perhaps, something like that. I can't get out of it. I'm not worrying about the Kingdom of Heaven; I'm worrying about the children and what is left for them. We are bringing them into a world that is poisoned, polluted, and dangerous. I feel so ashamed of what we have done, just in my lifetime, to this planet. Even the wild forests are threatened. I feel ashamed when I look into the little, shining eyes of my Roots & Shoots kids because they are so excited about what they have done, recycling, or making little organic gardens, or whatever they have been doing. They are changing the world, and they are excited.

You think of how it is going to hit them and you know, because I have seen it happen. Basically, when they get to high school or their first year at university, they realize the hugeness of it all and they feel they have to try and solve that. That isn't fair. So that is why Roots & Shoots is my driving force, really. That's what's keeping me going.

It seems that the most important thing for kids to know is that each individual is important in effecting change and that they should focus on this rather than the huge bulk of the Earth's problems. Focusing on the immense size of the problems can weigh them down.

JG: Exactly; that is what I am teaching them. Act locally; see that you are making a difference. Know that in seventy countries around the world, people are doing the same thing in their Roots & Shoots groups.

That is why I am passionately growing Roots & Shoots everywhere. That is really "my thing." What is the point? Okay, Conservation International is doing an amazing job. They've got this huge pot of money, and they could even buy up some of the forests, which probably isn't the solution, but at least it's a start. But what is the point of all these organizations sweating their blood to save a little piece of this and save a little piece of that if we are not raising new generations to be better stewards than we have been?

People ask me, "Have you abandoned conservation?" I tell them, "Absolutely not!" I am trying to make sure that everything I have struggled for will get carried on into the future. The only fear is that all these big efforts have started too late.

Maybe it's not too late. Maybe it's exactly the right time to make people really conscious of what's around them. When everything is fine, we are tempted to take it for granted, as we do in our youth. And now, because so much is disappearing, we realize how important everything is and we can start to love all these things much more.

JG: That indeed is true. But what I mean by "too late" is that we might have reached the point of no return, the point at which life on Earth, as we know it, cannot continue, the point at which there is more destruction than anything else.

The global climate shows us so much of that. You know that we have dolphins now in our English Channel, near Bournemouth. The birds have changed there, the butterflies have changed. It is a mixture of the warming temperatures and pesticides.

I remember Rachel Carson writing *Silent Spring* in the sixties. It made a big impact, but it didn't change anything. And it still doesn't. I won't see a point of no return, though. There is a chance now; there is a window. But that chance depends on enough people rolling up their sleeves and getting to work instead of just talking. It depends on enough people realizing that their actions, every day, indeed do make a difference—especially with what we do or don't purchase: More people need to be aware of the danger of certain buys.

But isn't what we think also important?

JG: Yes, and our attitude.

Jane, what is the place of humankind on Earth? You talk about "stewards," but isn't that anthropocentric? Shouldn't we also understand that we are part of a whole and reconsider our position accordingly?

JG: It may be anthropocentric. But either we have to be good stewards or we have to get off the Earth altogether! We are in a dominant position—because of our brains—and I would like to see us all go back to being like Native Americans or Pygmies, living in harmony with nature. But we know that won't happen. Therefore, the only option we have if we want to save the world is to be stewards. We must consciously control population increase, we must reduce the standard of living in the West, and we must consciously stop the developing world from making the same mistakes we have made in the developed world. We can make people understand the way of the West has brought disaster. Those are the three very, very tough propositions. But if we can't find ways to accomplish them, it will be too late.

You know that Edward Wilson* has said that if everyone in the entire developing world would obtain the same standard of living as the average Western person—particularly American—we would need four whole new planets to sustain us. This is the frightening thing now with China or India. China is on the brink.

---

*Edward O. Wilson is author of *The Future of Life* (New York: Alfred A. Knopf, 2002).

People are also talking about sustainable development at the Earth Summit. I don't like to talk about sustainable development, because development is the one thing that has caused all the problems. So I prefer to use the term *sustainable lifestyles*. That is all we need to aim for.

I share with Jane my concern about the misunderstanding of seeing ourselves as the center of the universe, which has brought us where we are now. Most people seem to have accepted this central position as a truth. Can we really make a change, then, not knowing and feeling that we are part of it all? If we feel separate and at the center, then we probably won't do what's necessary. We will act only if something touches us personally.

JG: I feel entirely part of nature in the forests, and I feel totally committed to trying to be the best steward I can. And that is the feeling that I want to give to all Roots & Shoots children: that they are a part of nature, that animals are their close kin—and that they are responsible, because they have the human brain, the language, the Internet. We can't undo all that. We can't go backward, because nothing ever does.

You aim to reach the children, but shouldn't we also aim to reach the politicians?

JG: I aim for the children because they are the next generation. But that means you have to aim at the parents and teachers, also. And the politicians? We all tend to blame the politicians, but who puts them there? Only if you have a sufficient groundswell of the voting public who is going to vote for the right politicians can you blame the politicians if they do it wrong.

Right now, it seems as though we are electing very conservative governments.

JG: Look at the Bush administration. It is the worst that can happen to the environment because everything is related now to the War on

Terrorism. So in the United States, many are afraid to show they care about the environment.

People ask, "With all the poverty and other problems in the world, how can you think of nature?" I am amazed by this question.

JG: If you don't think of nature, your situation is going to become worse and worse and worse, because the poorer people become, the more it seems their populations increase. We know that as women's education increases, family size drops. So the answer to the degradation of the environment by the poor is to improve the standard of living at the same time as offering family planning opportunities and educating women and providing health care. If you say "We must give those poor people more land" and then cut down the forest, the soil becomes increasingly fragile. Once-fertile soil becomes arid, desert spreads, plants change, poverty worsens, as do hunger, droughts, and flooding. So it is a whole spiral of which extreme poverty is an underlying cause.

And it is not only the forest that is destroyed; it is also the energy of the forest.

JG: Yes, because it is a nearly dead forest now. There is a possibility of life returning, but it will not be the same. Still, it can be a very beautiful forest again. There is no good having a black-and-white argument with loggers who say, "Well, we do sustainable logging." It isn't sustainable, in that certain tree species they are taking away will not come back—at least, not for thousands of years. But it is sustainable in that the forest has left animal species that can recolonize that forest.

I get upset when people stick to that black-and-white argument, because a lot of efforts by conservationists to try and reduce the illegal bush meat trade are being very much hampered, if not brought to an end, by strident activists. Compromises may be undesirable, but you won't make progress without them—as long as you do not compromise on your values. If you can't listen to the arguments of the others . . . That is a really important lesson for kids: that you must listen to the two sides of an argument. You must try to understand the perspective of the other side.

Do you know Professor Arne Naess?

JG: I don't know him, but I do know of him.

He is a Gandhian, and he talks about nonviolent communication. For him, that is clarity: saying what you truly think and feel. I believe your work is similar. You talk about compromise, but you don't compromise in what you say. You are very clear.

JG: I would love to stop the logging of the forests right now. But if I can't wave that wand, then at least we have to try to improve the situation as it is, while working hard toward a situation in which it will be different. But the one does not preclude the other—although then you are accused of jumping into bed with the enemy. That is ridiculous. Nothing changes if you don't talk to people.

Arne Naess says he doesn't want to be an environmentalist because that is putting himself too much in the center.

JG: People always say to me, "How shall I describe you?" And I always reply that I don't know. What am I? I don't know what I am anymore. I was an ethologist; before that I was a naturalist. Conservation education is something that I certainly do. I have never called myself an environmentalist. I don't know what I am and I don't care!

When we look at nature, we cannot escape looking at ourselves. We can't go on polluting and being in love with nature because nature invites us to live sincerely according to our deepest values and to be responsible for everything we do (or don't do) and for what we think, how we live, and how we treat ourselves and others . . . Changing the subject a bit, do you think there is awareness in all life-forms?

JG: I don't think I would call it awareness. That is a question that is getting close to *soul* and all of that. What is a soul? I suppose my view is more like that of the Native Americans, the indigenous people, where you have this great spiritual power. And I feel myself that within each living organism is a spark of that spiritual power. Of course, then there is rationalization and discussion, and we like to pinpoint things, clarify the way we think: That spark, we call it a soul. I think if I have a soul, I am quite sure my dog Rusty had a soul, too. And I expect to find him if I end up in some different form of being—which I believe I shall. I expect to find him there, too, also in some different form. I hope it is not too different, because he was perfect the way he was.

Once, when I was touching these huge forest leaves, it was like a cathedral. I think it was Gandhi, but I'm not sure, who said, "Why is it that when we destroy a man-made object, we call it vandalism, and when we destroy something made by God, we call it progress?" When I was looking at this huge tree, maybe nine hundred or a thousand years old, Mike Fay said, "Yes, and then the loggers come along and it takes thirty minutes to get it down. And they leave most of it lying on the forest floor." (She reflects quietly for a moment.) I don't know if the soul of a tree is the same as the soul of an animal, but it has a living spark, this piece of spiritual power.

I have asked Matthijs Schouten, "Do you think we are the last life-form on Earth in evolution?"

JG: I don't think that we as humans have reached the end of our evolution. But whether we will be the last? Probably that will be some creepy insect or a virus.

Today, people are offered such a variety and abundance of meat and nobody seems to think, "Let's reduce the number; we don't need all that food."

JG: I'm just doing a new book on vegetarianism. It's a mixture of . . . it's not really a cookbook, although it might appear on cookbook shelves. It will have some recipes, jewels that are coming to me from all over the world, from Roots & Shoots kids, from their grandmothers. So each recipe has a history. It is not the sort of book you would pick up if you want to start a vegetarian diet. It's the philosophy, the history. I'm trying to get recipes and illustrations from Gandhi. One of the illustrations will be of the yellow mushrooms we found in the Congo forest. They are the Pygmies' favorite, so we picked them and they cooked them up for us that same evening. I want a picture of that in the book.

And then, of course, I write about why vegetarianism is important. It's important for ethical reasons, health reasons.

Will Tarzan's favorite recipe be included in the book?

JG: (Laughing) He was a meat eater! But it is fascinating, the number of people who are vegetarians who have written about it, such as Leonardo da Vinci, Abraham Lincoln, so many people like that who have been very strong about it. Leonardo, for example, said that the bodies of meat eaters are living graves for the animals they consumed. It is beautiful, but very strong to say. I just say, "If you look at a piece of meat on your plate, what does it represent? Fear, pain, and death. Do you want to eat that?"

From what you have told us, it seems you need or want to be in constant contact with both nature and people in order to be learning all the time.

JG: Yes, we can learn from nature and people, and we can be shocked by them. But that goes for animals, too. They can shock us, because they can show brutal behavior as well. It isn't just we who show it.

But animals are brutal for survival, not out of cruelty.

JG: That is not true. The thing is, we act brutally knowing that we are causing suffering. I think the chimpanzees, to some extent, understand suffering—they definitely show empathy for each other—but they can't know to the extent that we can. Therefore, I would say that we are capable of evil and they are capable of only brutal behavior.

How can an animal be cruel if it doesn't think cruelly?

JG: We don't know that they don't think cruelly. I am so amazed by what chimps can think. I would never say that they couldn't think cruelly. I don't think they are capable of deliberately planning torture, for example; but they seem to enjoy inflicting pain. The young males certainly enjoy territorial combat. They seek it out.

Is that cruelty?

JG: It results in cruelty.

But it starts as survival.

JG: No, I don't think it does.

Does territory have to do with survival?

JG: To some extent. But when you see eight males attacking one stranger female when she is begging for help . . . They could gather her into their group and that would actually help their genetic survival, bringing in a stranger from outside. And they will take her when she is an attractive young adolescent. But if she already has a baby, they won't take her. And I don't think that has at all to do with survival . . .

Chimps are more interesting than other animals in that they can start something because perhaps it is adaptive. If a stranger female comes in and has a baby, it may be adaptive to kill her baby, because maybe it was sired before she tried to join the new group. So, they are perhaps killing the genes of an outsider. Although actually, they really need them. You see, the chimps can make this behavior into a habit. And they can continue to kill her babies, maybe two more, even if there is no way that they could have been sired by anybody but one of them because she is now part of their group. So, they seem to be in a strange gray zone, where instinct alone isn't enough.

If you take many animals and you raise them away from their kind, they can cope with many of life's problems . . . but they really need to learn. For example, for the females, it is very nonadaptive not to cut off and eat the placenta. Some female chimps [raised away from their kind] do, but others who, we think, didn't ever watch their mother do this cradle the placenta along with the baby, which is terribly nonadaptive, because if they jump and the placenta drops, it can pull the baby away. They haven't been taught, "Okay, we have a baby; let's eat the placenta."

I don't often say that chimps aren't capable of this or that. Because they are always surprising me. They are so like us, frighteningly like us.

And the gorilla is very different from the chimp?

JG: The gorilla is gentler. The bonobo has the same distance to us as the chimp; there is a little triangle, as far as DNA is concerned. They are gentler, friendlier, so most people like to choose the bonobo as the one that is closer to humans.

Chimps are generally okay toward people. It is more the interacting among one another. And hunting and killing for survival is generally acceptable. If I am stranded somewhere and there is no vegetable food to keep me alive, and certainly to keep my child alive, I need to kill an animal. I don't see that there is anything terribly wrong with that. An animal needs to die one day. A male super-hunter has made a successful hunt of a monkey and he is eating it. It is all his. He has no more need for food; he's got all he needed and can't eat more. He watches the others hunting, suddenly throws this dead baby monkey to his friend, who is begging, and goes off to start hunting again and kills another monkey. He is not [being] altruistic to his friend, I assure you! It is the excitement of the hunt, the excitement of killing.

Do you consider yourself to be a pioneer?

JG: I was with the chimps.

Did you ever feel lonely as a pioneer?

JG: I always had enough friends around. I am really lucky in that way. They are all over the world and do so much to help.

You said that when you started to work with the chimps, you wanted to get away from people and noise. Is there is a certain loneliness in your life?

JG: I love to be alone. I don't know why. But being alone isn't the same as being lonely. Being lonely is a terrible thing, but being alone is not. I think it has to do with the fact that I had a wonderful family. I had huge support as a child and was able to be very confident in myself. I had a mother who told me, "If you really want something and you work hard, taking advantage of opportunities, you will find a way. Never give up." She always praised us when we did our best and didn't expect us to do more than we could. Whereas my father, the few times I saw him, I remember telling him once with huge pride, when he came back on leave from the army, that I was second in my class. And he just said, "Why weren't you first?" And there are many parents like that. This is so bad. But in general, I was confident in myself, and that makes it much easier to be by yourself. I have always liked to be alone. That is why I climbed into that tree as a child. I had the kind of childhood every child should have.

Have you met with strong cynicism, or was it expressed only in the beginning of your research?

JG: I still do, but mostly from people who are invasive—the intensive farmers, the hunters, and the trappers. They can't, can they, admit that animals have feelings? So they blinker themselves, and they become very schizophrenic. The scientist who puts on his white coat and goes and tortures dogs in the lab, then comes home, without his white coat, and says, "My dog understands everything I say"—for me, that shows that there is a split in the brain somewhere. And it must be terrible for them.

What is love?

JG: Oh, dear. (Long silence) I can't possibly answer that question without contemplating it for several months! Can animals show love? Yes. Can a tree show love? No. So love has to be active. Love means accepting faults and still caring about that person. I suppose it is based on the kind of love that a mother has for her child. I think I see that as the root of love, because that's a love that embraces everything. It embraces faults, physical and psychological, and it is giving and caring. From that you get your different degrees of love.

I have said in one of my books that we use the word *love* so sloppily. We love God, we love our country, our child, our husband or wife, and we love ice cream. We use the same word for all of that; but in each case, it means something different. So, it is hard to define a word that has so many meanings. But as a sort of core meaning, I would say that love is the mother's love for a child and the child's love for its mother. It is real, countable, and it has a human face to it. Next, I would put the love you can have for a dog and the love the dog has for you. It's funny: When people ask me what my favorite animal is, I'm supposed to say a chimpanzee. But I don't! It's a dog. I just love dogs.

Gareth Patterson is in touch with lions in a very emotional way. They come and fetch him when they have their babies. There is a love, a connection there. But I have the feeling, though I don't know if it's right, that you are a bit more distant from the chimps.

JG: I guess that's because I love observing. But with a dog, I am not observing anymore; I am *being with*. I really want to look into the world of the chimp; I don't want to be part of it. I want to understand what it is like to be a chimpanzee in the chimpanzee world. I would give anything to know what it is like to be in a chimpanzee mind and to look out through chimpanzee eyes. I would love to know that. But I never will. That is part of the wonder. Whereas, if I'm with my dog, it's nothing like that. Then it's just complete togetherness.

Why this distinction?

JG: When you have a dog that you share your life with, it isn't a wild creature living on its own that you don't want to interfere with because you want to leave it as uncontaminated by us as possible. The dog, inevitably, is part of our world.

I remember someone visiting me in Dar es Salaam. The dog I had at that time had a horrible thing where her eye just blew up. I was completely hysterical about it, didn't know what to do. But I found her a vet who knew what it was, and this man, in a rather discouraging way, said to me, "Well, if you're like this over a dog, I can't imagine what you're like when one of your chimps is hurt." And I said, "It's nothing like the same thing, because the dog looks up to you as kind of a goddess, totally dependent on you for everything, and chimpanzees are wild and free and they are not dependent on you. If I can't help them—which I try to do—it isn't a betrayal of trust." So, it is about trust. I think I have had that from my dogs as much as they from me. I trust that they are not going to bite me and that when I come back after an absence, they will still love me and understand that I didn't want to leave them.

That trust relationship with the chimps is very different. They trust that I am not going to hurt them, but they don't expect me to provide for them and they don't expect me to make them well—or even to be there.

Does that mean that we never really understand each other unless we really share our lives?

JG: No, I think we can understand pretty well. Looking through that window into the world of the chimps, after all forty-three years, I think we have a pretty good understanding what it is to be a chimp in the chimp's world. And some of the chimps I do love, but I know they don't love me. If you have a captive one and you treat it as part of your family, then you get the same kind of love as with a dog, except that chimps are not trustworthy. They are like some dogs that aren't trustworthy.

Did a dog ever betray you?

JG: No.

And people have?

JG: Oh yes, of course.

Works by Jane Goodall

Africa in My Blood. Edited by Dale Peterson. New York: Houghton Mifflin, 2000.

Beyond Innocence: An Autobiography in Letters, The Later Years. Edited by Dale Peterson. New York: Houghton Mifflin, 2001.

Brutal Kinship (with Michael Nichols). New York: Aperture Foundation, 1999.

The Chimpanzees of Gombe: Patterns of Behavior. Boston: Belknap Press/Harvard University Press, 1986.

Forty Years at Gombe. New York: Stewart, Tabori, and Chang, 1999.

In the Shadow of Man. Boston: Houghton Mifflin, 1971.

Innocent Killers (with H. van Lawick). Boston: Houghton Mifflin, 1970.
My Friends the Wild Chimpanzees. Washington, D.C.: National Geographic Society, 1967.
Reason for Hope: A Spiritual Journey (with Phillip Berman). New York: Warner, 1999.
The Ten Trusts: What We Must Do to Care for the Animals We Love (with Marc Bekoff). San Francisco: HarperSanFrancisco, 2002.
Through a Window: Thirty Years Observing the Gombe Chimpanzees. Boston: Houghton Mifflin, 1990.
Visions of Caliban (Jane Goodall and Dale Peterson, Ph.D.). Boston: Houghton Mifflin, 1993.

Jane Goodall has also published several children's books and many scientific articles. Web site: www.janegoodall.org

ARNE NAESS

Being in love is the most
wonderful form of madness.

Stepping out of the plane with Jessica in Oslo, Norway, three words come to my mind: *wood, clean, space.* The airport building is finished in wood and it smells delicious as we walk through it. It is surprisingly clean, as are the station and the train. Here people seem to care for what they use and how they use it. And there is space!

Modest wooden houses are set on short rolling hills covered with birches and pine trees. Low clouds cling to the land as the train races us to the capital, where we will meet the philosopher, author, and mountain climber Arne Naess. I was interested in exploring with him his consciously chosen path, which has resulted in a long and active life. Naess has always done what he feels he needs to do, and whenever he can, he will choose the unexpected. His clarity and humor are a joy.

Born in 1912 near Oslo, Naess studied philosophy, mathematics, and astronomy at the University of Oslo, followed by studies at the Sorbonne and the University of Vienna. For his postdoctoral research, he attended Berkeley. At the age of twenty-seven, he occupied the University of Oslo's chair of philosophy, which he held until 1969. In that year, he opted for an early retirement to dedicate his time to ecological issues. This retirement resulted in the publication, in 1976, of

Naess's classic book *Økologi, Samfunn og Livsstil* (Ecology, Community, and Lifestyle). The book was later translated into several languages. The book and his analysis of environmental conflicts have earned Naess a substantial reputation within the international environmental movement.

In the philosophical realm, Naess has had a strong influence on the development of philosophy in his country. During the first fifteen years of his work at the University of Oslo, he was the only professor of philosophy in Norway. He helped to shape the Norwegian intellectual environment in the post–World War II era and was the man behind the development of the Oslo school of philosophers. The man we will meet is considered to be one of the most important contributors to philosophical thought in the twentieth century, and even at the age of ninety, he is a frequent speaker all over the world. The founder of *deep ecology,* he has given his name to the Arne Naess Center for Ecology and Philosophy (ANCEP). Naess's work has earned him several awards, including the Mahatma Gandhi Prize for Nonviolent Peace (1994), the Nordic Prize from the Swedish Academy (1996), and, earlier, the Sonning Prize for his contribution to European culture (1977). This last award is a Danish award similar to the Nobel Prize, given for contributions to European culture. In 2002, Naess received the Nordic Council Award for Nature and Environment.

He has been active in other arenas, as well. One of his primary interests has been mountain climbing, and he headed the first Norwegian expedition to the Himalayas. Only recently has nature made him aware of the limitations that come with age.

What is the balance between the philosopher and the man, I want to know, given that nature is such an integral aspect of the man. He is conscious that nothing exists on its own. He has learned to be connected to nature, as he is to his fellow human beings. The separation between humans and nature has become inconceivable. He lives this unity, thinks this unity, sees it, smells it, and feels it. As we learn in our conversation, he is a Gandhian at heart who practices nonviolent communication in his daily life.

At this stage, he can no longer live in his mountain cabin as much as he would like. He has spent a great deal of time in the vast open territory above the tree line in the rugged Norwegian mountains, surrounded by nature and silent hills. With books as his true companions and no need for electric power or heated water, he has lived with nature in its purest form. For the occasional guest hardy enough to risk the hazardous three-hour climb to the cabin, Naess has built a guest room because, as he puts it, "Guests talk." He named his cabin Tvergastein (meaning "a heap of stones") after the outcroppings of stone that surround it. He has also honored his retreat by naming his theory of ecosophy—*ecosophy T*—after the place where he has done most of his thinking, acknowledging the importance of the mountain and his cabin.

Our taxi finds its way to his city house, a dark wooden structure in a row of four houses. We ring the bell and his young wife, Kit-Fai Naess, opens the door to welcome us. Arne stands in the room behind her, curiosity shining from his clear eyes. He takes us in, every detail, in one sharp look. Did we pass the exam? I don't know, but our subsequent four-hour conversation suggests we did.

The house is modest and warm, with wood everywhere: the walls, the ceiling, the furniture. He tells us how important it is for him to be surrounded by wood—the texture, smell, and feel of it. I am in awe of the enormous energy this man emanates. At ninety, he seems filled with as much zest for the essence of thought, for the love of life itself, as he must have been in his early years.

In each of his answers to my questions, I notice a moment of hesitation. Only later, near the end of our conversation, do I realize this originates from a conflict within him between the philosopher and the man.

As Jessica and I situate ourselves with tape recorder and notebooks, Naess looks at our paraphernalia and feigns amazement. I go along with the act:

I have all these questions for you!

Arne Naess: Questions?

What were you expecting?

AN: (Joking) Some descriptions of bad manners, maybe?

His little joke sets the tone for our talk. He seems at ease and hasn't bothered to change out of his comfortable slippers. More than anything, his eyes tell us that the years have not taken away his sharp mind. He is a playful, witty, mischievous, wise man, ready to answer and react to whatever arises in our dialogue today.

I tell him how much I was inspired by his books, and how his independent thinking made me laugh out loud at times. It was fascinating to meet him through his writing. Reading between the lines, his work left me with the sense that he is the kind of man who does exactly as he pleases. "He does!" says his wife, who joins us during the first part of our conversation. We are talking to a philosopher pure sang, *and a sometimes rebellious scientist.*

What is nature?

AN: The meaning of that word is different on different days and occasions. It is nothing definite, something that is changing in different contexts. That is because it is a word, not a concept.

But what does it mean to you?

AN: For me, it is a word with interesting, different meanings. I like to study the development of the uses of such a word. First, there is *natura,* a Latin word used in sentences like "What is the nature of so and so? What is the nature of the illness?" And then you have a whole list of different uses.

But does it have a specific meaning for you?

AN: For me it is this list of different meanings. That is nature.

Does it have an emotional meaning?

AN: That depends completely on the context. If the context is the Norwegian nature, it means tremendously much to me—a feeling that started in my youth. There is this picture, when I was only three years old, with the family. You see me running into the woods. I guess it was the absence of people that drew me there, like the cottage where I have a few square miles around me with no people. I look out of the windows and think, "No humans!"—although I am not particularly an antihuman person. Next to the cabin are some vertical rocks that almost fall into the house. Some people warn me, but for me it is okay if they take over.

Matthijs Schouten tells the story about a convent in Tibet that was overtaken by mice. Although the nuns initially tried to get them out, the mice made it clear they were very much in the right place. So the nuns decided to leave the building to the mice and moved elsewhere.

AN: That sounds very similar to my eight points of deep ecology.

As the founder of the theory of deep ecology, Naess includes in his book Ecology, Community, and Lifestyle *the eight points that describe the ideas behind it:*

1. *The flourishing of human and nonhuman life on Earth has intrinsic value. The value of nonhuman life-forms is independent of the usefulness these may have for narrow human purposes.*

2. *Richness and diversity of life-forms are values in themselves and contribute to the flourishing of human and nonhuman life on Earth.*

3. *Humans have no right to reduce this richness and diversity except to satisfy vital needs.*

4. *Present human interference with the nonhuman world is excessive, and the situation is rapidly worsening.*

5. *The flourishing of human life and cultures is compatible with a substantial decrease of the human population. The flourishing of nonhuman life requires such a decrease.*

6. *Significant change of life conditions for the better requires a change in policies. These affect basic economic, technological, and ideological structures.*

7. *The ideological change is mainly that of appreciating life quality (dwelling in situations of intrinsic value) rather than adhering to a high standard of living. There will be a profound awareness of the difference between big and great.*

8. *Those who subscribe to the foregoing points have an obligation, directly or indirectly, to participate in the attempt to implement the necessary changes.*

AN: The basic thought is that every living being has a value and a meaning in itself, a meaning different from being something good to eat or to use. In other words, it makes sense to assist all living beings. This has had a great influence, also, while I was in my cottage: that no living being should be disturbed. That was difficult at times, to take care of everything within a circle of ten meters. We have to protect and not trap things—not even the grass, if it isn't necessary. For example, if there are bricks, it is better to step there.

How does nature inspire you?

AN: Near my cabin is a great mountain. I have the feeling that this mountain does not dislike me; actually, it rather likes me. I have this very close way of looking at it. During terrible storms, it makes the place even more special for me. One time, when I was sitting inside the cabin, three quarters of the roof went sailing off in such a storm. I had to rebuild the roof, but it was a very good thing. It showed us that there is something stronger! Much more important, if you love nature, is to study nature.

Could you say it is a dialogue with nature when the roof is taken by the wind?

AN: (With profound conviction) Oh, yes! It [nature] also talks back to me! It is an enormous test, being part of this dialogue, just sitting down, doing nothing for sixty minutes but listen to what the trees are saying. What I mean is that the trees express something, be it tenderness, depression, or a certain power. It is good to learn their language and treat them as living beings. I believe it is very important to get small children—even as young as four years old—out into nature. We have to bend all the way down to show them the small details of nature.

Did your parents show you?

AN: Oh, no! My father was an authoritarian man. He died when I was only one year old, and I guess that was a blessing for me. My mother acted like a martyr. But she never said no, so I could do what I wanted. And I went outside, because that felt like what I needed most.

What is your nature?

AN: (Making a gesture with his arm to indicate *straightforward*) Like this. Trying not to hurt others. And at the same time, rather often, you have to say the honest truth. But what you have to say should correspond with the proper body language; otherwise they won't believe you. (Shifting from a sideways position on the sofa to facing straight ahead) Even if they are on the wrong side, you have to approach your audience as fellow humans.

That is the attitude Nelson Mandela had when he came out of prison.

AN: If you get near a person like that, you have to ask yourself: What would I have done with an upbringing like that? You never know, do you? Everyone will open up, if you have the right body language and if you are honest. I go my own way, but with complete respect for my fellow humans, whether they are murderers or not. I have been asked to give lectures in prison, and I did. I said, "You are not a murderer; you have committed murder." They looked completely ordinary, but had done terrible things.

You say that you don't see those prisoners as murderers, but rather as people who have murdered. Would that be any different if they had murdered one of your loved ones?

AN: My feelings would be terrible, of course. But as a philosopher, I probably would still see them as fellow human beings.

Is yours merely a philosophy or is it also a deep feeling?

AN: Yes, in the long run, it becomes a deep feeling.

Does the fact that you try to live your life showing respect toward all human and nonhuman beings mean that you expect others to do the same?

AN: *Respect* is a very strong word. I box in Gandhian style: only for fun, for the repetition of the movement.* (Laughing) But my teacher was afraid to hurt an old man.

Boxing has taught me to be fast, because if you open up for a moment, they [your opponent] will hit you."

So it teaches you not to be open?

AN: Rather, it teaches me to react fast!

But how does this correspond to your respect for all life? Doesn't this philosophy of living require an open attitude?

AN: Yes, of course. There is not much to hide if you are not afraid of your shortcomings. Respect for a human being should be independent of what that person has been doing in his or her life.

Do you think it is easier to love and respect others if you know and like yourself?

AN: If you don't like yourself, it is very easy to dislike others and nature. That is why your upbringing is very important.

Did your mother raise you to like yourself?

AN: I saw my mother's dislikes, but gradually did my own thing, whether

*Until recently, Naess boxed at the University of Physical Education and Sports in Oslo. A few years ago, two Poles—Nina Witoszek and Piotr Kuzinski—decided to preserve Naess's scrimmages in the boxing ring for future generations by making a short documentary featuring him and this sport. The result can be seen on the Internet at: www.sum.uio.no/staff/arnena/boxing/index.html.

she liked it or not. I started explaining to her that whatever I did was not as dangerous as she thought—climbing mountains, for instance, can be done in a way that makes it less dangerous than driving a car.

As a child, you told us, you ran into the woods. You said it was the lack of people there. But there must have been something more that attracted you.

AN: I ran away from my family! I found small things that attract every small child. (Responding to his wife's comment from the other room: "Pine cones.") Yes, pine cones and things like that. I looked at the trees in the woods as human beings: some strong, some old, others weak even when they were not that old. Many times a year I went back and looked at each individual tree and saw different "human beings," alive. It was strange. I haven't seen other people looking so intensely at trees. All the trees are really different to me.

How did it feel for you, to see those trees and to be in the forest?

AN: I felt so near the forest. It felt as though I was next to human beings.

But you didn't like people?

AN: No, I was running away from the family, from their rules: You shouldn't do this; you can't do that. Without those rules, I somehow felt I could be me, accepted there in nature.

You wrote in one of your books: "The mountain is god, and perfection."

AN: I say that for me, a mountain is what God is for others. And if you like to "absorb" trees and mountains, as I do, you realize that some are godlike, others are smaller.

Is there a sacredness in every one of them?

AN: Yes, and to destroy nature is to destroy the sacred. How can you do that? For me, it destroys something in myself, because of my identification with nature. It hurts me when nature is destroyed.

How do you define deep ecology now?

He asks Kit-Fai to hand him his eight points and starts laughing as he rereads them: "Did I change any of them?" he asks her.

She tells us that he had altered them somewhat when he received so much criticism for his statement that "a diminishing of the population will be good for both humans and nonhumans." She explains that her husband was called "antihuman" and even a fascist.

AN: Yes, I said it would be good for humans if there were fewer of us, and very good for all the other living beings as well.

But it is true. People think most about themselves and very little about the effect overpopulation has on other life-forms. It is even more true now than when you first wrote it.

AN: Of course. If everyone would have no more than two children, it would be fantastic. Then the total number of people on the Earth would be reduced.

His wife adds, "Yes, but in the beginning, you forgot to mention that you were thinking in the long term."

AN: Of course. I wasn't thinking of killing those who are already here!

I remember the reaction one of my sons had as a very young boy when I read a newspaper article to him about elephants that had to be killed because there wasn't enough space for them. He asked, "Why kill the elephants?" A point you often make is that everybody should find something that he or she considers worth fighting for in order to realize the goal of a sustainable world. What do you consider worth fighting for?

AN: Teachers should teach young children in the early stages of their lives to have respect for nature. That sets the tone for the years to come. For instance, if you look at the industrialization of our country, you see that the Norwegian government wants at least one industry here that we all know is bad for nature. The people have a great esteem for nature, but politics are different. Politicians don't act at the same level of feeling as do people. People ask themselves, "How can this be?" Norwegian people like to go outdoors. The labor party has actually included a statement in its political program that every family should have access to a cottage, but it was left out later. I thought it was a very nice idea; it showed that they saw the importance of having a place where you can "restore" during the weekend. But why do politicians fail to feel like other people?

It's similar to what Rupert Sheldrake has discussed with us. He spoke about how people live a kind of duality: During the week, they work very hard in a job that is bad for nature, and then on the weekend, they spend time in nature to be restored for the week to come and to have the energy to work on issues that will destroy nature.

AN: We are certainly not consistent! In Norway—even in the north—people go to their cottages to be away from the pressure. When they come back, they like human beings more than before.

Well, at least that's a form of continuing the love of nature! How often do you go to your cottage?

AN: Until recently, I went three or four times a year, at least two to three months of the year. To get there takes as long as to get to Italy, but it is worth it. I stopped going there last Christmas. The storms are such that you don't know what to expect. This last time, it was simply too cold for me—which means that it wasn't romantic any more! I just sat there in the room, too cold to be outside. After five minutes outdoors, you would have to run inside to hide from the cold and the storms. And it was dark after three in the afternoon. There is no electricity, only some kerosene lamps. I can't enjoy all that as before. But I still consider those fifty thousand square miles [the Norwegian mountains as a whole] to be an overwhelming greatness. It makes you aware of your own smallness.

Naess once explained that with all this nature in front of you, it is impossible to thinking polemically, to fashion a polemic against other philosophers. For that reason, the cabin was a good place to be working on his philosophical projects. It offered the holistic influence of nature at its best. I am curious to hear some more about this influence.

What has nature taught you most?

AN: What greatness is. If you want to be a great man or human being, it should be in a way that is right with nature. I never had a professor-like attitude, but instead more a student attitude, so I could also learn from the students. They even taught me how to interact with female students. I was married very early because that was what my mom and grandmother wanted, but I ended the marriage after ten years. I felt the need to be clear about myself. That is part of who I am.

We agree on this. Clarity is a very strong part of my philosophy, too. It has everything to do with who we are in nature. The clarity in you is the same clarity that facilitates contact with nature. If we have "polluted" thoughts, it is more difficult to feel the contact.

AN: And we have polluted thoughts! But it is better to admit that and then move on. We always have excuses. I am ninety years old, for example. It is all a matter of facing your mistakes. Admit what you did wrong and right. Looking back, I think I did many good things in my life.

Regarding overpopulation, how do we find the balance between human needs and the needs of other living creatures or life-forms?

AN: First of all, parents have an important role in pointing out to children that you have to respect other living creatures. Take, for instance, the snake: It can be dangerous to us, but there should be areas where snakes can be left alone. Every life-form should have a space to be.

But if you look at the map of the world, you see that there are so few protected areas. How do we find the right balance?

AN: We can offer assistance to developing countries—people who, instead of serving in the army, go to these so-called developing countries to serve eighteen to twenty-four months. They go there in larger groups to work and help out and do marvelous things; and they should behave properly, like guests, because it is not their country. They help the local people by digging wells, setting up solar panels, teaching them elementary hygiene—in short, doing what needs to be done in that local community while living as the locals do. They should make sure not to flaunt their Western way of living and consumption—avoid taking expensive photography equipment, for example, so as not to give the impression that things are so much "better" in the West.

If you look at all the topics of the 2002 Johannesburg Earth Conference—AIDS, poverty, health care, water, housing, the economy, the Earth—you ask yourself, "How can we balance all of these concerns without humans continuing to dominate everything?"

AN: Again, a great deal comes down to the parents; they are the ones who need more education. (With a mix of seriousness and humor) They should go to seminars to learn more. It would be good to make a law that says you can't be a parent without attending courses first.

Jane Goodall shares your ideas in this respect. Her program Roots & Shoots is based on similar principles.

AN: I know that you have to start educating children at the age of four. Once they are formed, it is very difficult to change and teach them.

The world population reached six billion in 1999 and grows by 1.4 percent annually—two hundred thousand more people every day! Where do you see the world on the ecological level? Where are we?

AN: This century we are at a very low level of ecological understanding and ecological behavior. It is going to get worse first and then maybe better in the next century. But that is still very far away, and for now, we are increasing harmful choices, like letting the world population grow. I disagree with the theory of a complete breakdown, but it will take generations before we learn to change. It is said that the new generations have set "getting rich" as their major goal. They have so many opportunities. I have to admit that I am a bad example, because I recently went to Bali to attend a seminar that ANCEP (Arne Naess Center of Ecology and Philosophy) arranged, and every day for two weeks I gave a lecture addressing varying themes around coexistence.

Naess went to Bali for, in addition, the opening of the Arne Naess Center, where students study various topics linked to nature issues.

AN: There is much to see in the one square mile around my house, so I really have not such a need to travel so much. We don't need to go to other countries or even leave the area where we live to enjoy the incredible diversity of nature. Each square meter has its own flowers, creatures, and infinite richness.

Some people say there's so much poverty and illness in the world that talking and thinking about nature is a luxury we can't afford.

AN: (Laughing ironically) I agree that we have to set priorities, but we also have a great responsibility toward our fellow humans. That is something very important, and nature is part of that.

Regarding our need for nature, an exception is made for the very sick and the poor. Why would they not need nature?

AN: Yes, all people do, even criminals. When you take them out into nature, they feel a tremendous relief. I have taken them out—literally—and they said that they want that more often, because they believe it would change them.

What happens when these people experience nature?

AN: Nature shows them how much things are alive. But you have to go with them to explain that they are all part of that nature. Without someone accompanying them, they would probably go straight to that neighborhood that they know to plan their next crime. But if you take them on long trips, it opens them up.

How did you open people up to all life? How did you get through to them to show them that they are part of nature's web?

AN: I stopped them every ten meters, showing them how to avoid walking on the grass and so forth. It builds their self-esteem again, looking at other species and realizing that we are not less, not more, and that we have our own little space here on Earth.

Jane Goodall makes a distinction between* nature *and the* natural world. *The latter is everything that we have not intervened in yet.

AN: That is an excellent distinction. But what about the sea, the mountains? The *natural world* is a very comprehensive term.

It is not a completely clear distinction. For example, pollution caused by our hand eventually reaches the sea, the woods, and everything else.

AN: And sometimes, in order to protect life in the woods, you have to cut many trees.

Sometimes you have to kill animals so that there will be enough food for their population and in the woods we sometimes need to cut down trees to allow light for others.

AN: Yes, it is very complicated. It should be acknowledged that there are different opinions. Farmers have to kill weeds, doing no more than what is necessary. But they still have to intervene a lot in nature. It is nice to hear modern farmers say that they see the beauty of nature. They are educated in a such a way that it is impossible for them to ignore this reality.

How does a philosopher look at the notion of awareness in nonhumans?

AN: Yes, awareness. I think nonhumans possess awareness. Animals are known to change direction all of a sudden, which means they must have an awareness of danger.

What about plants and trees?

AN: No, I don't think they have an awareness. But in Indian philosophy, they do . . . If I look at the flowers in my vase some days, they seem so dreadful that I am tempted to put them in a different room. I can't help thinking that they are in a certain mood. But philosophically, I don't think they have moods.

(Moving the pot of chrysanthemums that sits on the table) I do go pretty far in behaving as though they have feelings, though. These are looking extremely happy, exceptionally joyful! These feelings are realistic in the sense that if I placed the flowers in another room, I would do so because I feel sorry for them . . . But philosophically, I don't think they have a nervous system.

Do stones have awareness or feelings?

AN: Philosophically, I will not say they have feelings.

But you said earlier that the mountains and the trees talk back to you . . .

AN: (Struggling, as if conflicted) There is a clear distinction between my feelings about these things and the philosophical approach.

Once I was on sleeping alone on a mountain pass at full moon. When I woke up in the middle of the night, the moon stood high and flooded everything in her golden light. I looked at the huge rocks on either side of me, where I had placed myself for protection . . . and they seemed alive! I had to shake my head to see if I was seeing correctly. The whole landscape was alive. I was almost embarrassed to look at the rocks because it was as if they were making love to the moon. There was so much communication between them. Of course, what I saw was energy emanating from the rocks, caused by the full moon. What do you say to this?

AN: What do I say? I think your experience is being a human being! I think it would be completely irrational if you weren't open to that energy and feeling that joy.

There is an interaction between us and flowers, or water, or rocks, or the sun. What is it that comes from the other side?

AN: Enough for us to make us feel like that!

He makes me laugh out loud with his answer—because he just leaves it at that, while for me, this communication is a world of fascination. This leads to my next question:

Do we learn through science or through experience?

AN: Oh, you can go back and forth from one to the other. But we need our natural feelings more than we need science. *Experience* is our natural feelings, and that is what we need more in life.

When I asked Rupert Sheldrake whether nonhuman life-forms have awareness, he said that he would use the word* soul, *rather than* awareness. *He talks about the sun, for instance, as having a soul.

AN: Well, during the past one hundred years, scientists have given up on talking about the soul of the sun . . . I would say that in certain senses we have a soul; in others, not. That is all right for me. But we don't have a metaphysical kind of strange, nonmaterial soul.

Is the soul something that goes on eternally?

AN: Yes, and some people are less soulful than others, in the sense that they tend to act without a soul toward others.

(Grabbing a cup from the table) I see this cup as a living being. This is the nose, and it looks like a fat being to me.

You compare it to a living being, but has it got life in it?

AN: Not scientifically, of course, but for me, it certainly has life in it. This cup has a good temper. It is enjoying itself. (Putting down the cup again) It is interesting: With my death near, I read articles and think about theories and realize now more than ever how excessively anthropocentric we are when looking at things. We do not understand, for instance, that this cup in a sense is looking back at us, too!

Although we see everything from your anthropocentric point of view, we can realize that nothing is separate. Some people feel so separated that they are moved to kill themselves.

AN: Yes, how could they? They are not alone.

We pause before I ask my next question:

Do you think that deep ecology will disappear with you?

AN: From let's say 1969 to 1979, that was a fear, or, better said, a concern. I was invited to various places and so many people know about the concept that it has become a living thing now. It was important to get it articulated, stated in words.

It is interesting to me that through my own experiences, I have come to the same conclusions that you state in your theory. In your words I find a whole philosophy written down and feel I'm not alone in my way of perceiving life. That helps! So I think it is important that these concepts are articulated. It gives support to all those people who experience the oneness of life and revalue their place in nature. You say that in philosophy, there are not too many followers, that people can develop their own thinking. I like that.

AN: I didn't know you had come to the same thoughts, because we have never met or spoken before.

What is more important now, at your age: the philosophy or the movement of deep ecology?

AN: The movement, of course, because it goes its own way now. In a sense, it is completely out of my power. But naturally, I like to hear about it; I like it when people do something with it.

Because it benefits people or the Earth?

AN: Especially for the Earth, but also for the people. They can now talk about it.

How would you define your relationship with nature now, at this stage of your life?

AN: I feel bad that because of my physical shape, so many things are closed to me now. I cannot go here, cannot go there. I miss my cottage most of all. I would define some things as impossible for me, and that is hard sometimes. But I also feel that nature understands that I am very old, that I can't climb trees anymore. Nature has a friendly look for me.

Because you have been good to her?

AN: No, I wouldn't say I have been good to nature. But friendly, yes.

Do you consider yourself a pioneer?

AN: Yes, in some ways. I have liked it. It has been a strength to be more a pioneer than one who is elaborating and trying to give the finishing touch to things. I was more often the one who pointed things out, although there were people more clever than I was. But I think I needed things to go farther than I was able to go.

Have you experienced the loneliness that sometimes goes hand in hand with being a pioneer? And how have you coped with the resistance that so often comes with bringing out new ideas?

AN: I always had "critical" people around. In the beginning, I would take criticism very seriously. But it is the way it should be: They should say what they think, too. That is fine with me, as long as critics have seriously studied my work. I will never feel sorry about being criticized, because that's how it is. I can't have copies of myself around who completely agree with me! I have always moved on to new subjects, which makes me difficult to copy, anyway. My publications are more expressions of my thoughts. Whenever I felt that I didn't want to work on one anymore, I would publish in order to get reactions and start discussions. You can't get discussions without putting something into print.

This may explain why Naess has been—and still is—so remarkably productive. His work consists of more than thirty books and a long list of published articles. Even after his "retirement," he has not given up writing. His publications include an interesting assignment with the United Nations Educational, Scientific, and Cultural Organization (UNESCO). Based on the work of the Oslo Group (the school of philosophy that Naess has helped to create), he started to examine the nature of ideological controversies between the East and the West. Naess was appointed scientific leader of the investigation into the issues surrounding the Cold War. The Oslo Group became known for its empirical methods, using carefully constructed questionnaires to gather data. In this case, questionnaires helped to explore the different uses of the term democracy *down through the ages. The group surveyed approximately 450 people, mostly professors in various disciplines who lived in the East as well as in the West. They interviewed both Marxists and anti-Marxists, and then sent the Marxists' answers to the anti-Marxists for comment, and vice versa. The impact wasn't that beneficial, thinks Naess, for both groups continued their separate ways.*

AN: Apparently, UNESCO thought that talking about democracy was dangerous and that it would be better not to print any more copies of our document. But you should still be able to find it somewhere in the UNESCO files. The volume of results we printed was sold out, but has never been reprinted.

At this point, his wife invites us to have lunch in a room downstairs, offering her husband a well-deserved break from our discussion. But before we pause, I have one more question for him:

What is love?

AN: What is love? That is a very important question for me. "Being in love" and "to fall in love" are two very different expressions in Norwegian! Being in love is the most wonderful form of madness. But you might wake up one Monday morning to find out that there is nothing left anymore. Normally, love is something that goes on for the rest of your life. It should be a living kind of feeling. And I agree with the Norwegian language that "being in love" and "I love so-and-so" are very different. "I love you" should not be easily said.

Suddenly, we hear a loud "Ouch!" from Kit-Fai in the other room. And then: "The bread is burned!" This is meant to make it clear that it's time for our ninety-year-old conversation partner to take a break. The food does us all good, but it is Naess himself who asks us if we have any questions.

I would like to ask you the same question one more time—What is love?—because it is a central theme in this book and I would like to hear what else you have to say.

AN: I think that *love* is a very general name for very positive relations. To take care, whether it is of a stone, a person, or anything else, means that you will be unhappy seeing something go wrong with that thing or person. That is, at least, the philosophical part of it. But if you take twenty or thirty occurrences of the term *love* in newspapers, you will see it as a much narrower concept. As a philosopher, I am very interested in general attitudes: You love what you think is alive—and even if you are not correct about it being alive, you will have a different attitude toward it from the very beginning. So much is going on in the world because some living beings are not looked upon with mercy and concern, which is the way everything should be approached.

Does this come from a lack of love, or are many people unaware that all life-forms are actually alive?

AN: For me, it is in a sense independent of whether some living things can feel or not. For instance, I don't think a plant can feel anything, but I act as if it could.

But you mentioned earlier the idea of a certain awareness . . . To me love is the light, the spark that is in all life-forms.

AN: Well, that is a very nice way to look at it.

But you can't follow me?

AN: Oh, yes. And I think I like your philosophy!

But . . . ?

AN: Well, I have my difficulties in writing it down. I see problems ahead—for instance, that the term *love* will be misunderstood: "How can you treat a lovable thing like that?" Or: "How could you throw away living things?" But I like your words. I will never argue against them. Really, I congratulate you on your worldview, and I hope you continue on your way.

I can't help but feel the contradiction in you. I hear you making a lot of distinctions between your own feelings and what is philosophically possible. Am I right?

AN: Oh, yes. (Repeating, as if underlining the words) Oh, yes.

Do you regret being a philosopher and a scientist, from time to time?

AN: (With surprising force) Oh, yes, from time to time I would have been much happier if I had been a paleontologist fossil hunting and all that. I like that. In fact, I know there are other fields that would have made me really happy. With philosophy, you can never stop. You have to keep developing.

It is a bit of a harness, too?

AN: Sometimes as a philosopher, I cannot go to a certain place in thought, although the idea can have my complete sympathy. (Pause) No, it was a very bad choice, in terms of my happiness. And also, as a philosopher, I was trying to find my own way. But it was a very hard way to go.

Would you become a philosopher a second time . . . ?

AN: (Interrupting) No, absolutely not!

We take a little time to let Naess's confession sink in before we continue.

How are love and deep ecology connected?

AN: In deep ecology, when you see something worthy of your concern and you are able to satisfy its needs, that is a great thing, because every living being is worthwhile.

In everything that I have read of you or about you, you have never used the term* stewardship *in reference to humans' relationship to the Earth. Yet that is how the majority of us see our role on Earth.

AN: It is too much for us, to be stewards.

I personally don't like the concept that much because I feel that we are no more than part of the Earth.

AN: I agree completely. So far, we have been unfortunate for the Earth. (Ironically) But within twenty-eight thousand years, we may be its saviors, because of [the arrival of] some comet. We can calculate that it might fall through the atmosphere and end every human being. So in a very long run, we may save the Earth!

Jane Goodall says we are stewards because we are the only species that, to put it in my own words, "thinks about thinking"—that because there are so many of us, we are capable of doing great good or great harm.

AN: I think *steward* is a good word, but not good enough because this makes it seem as though we have a knowledge of what is going on here that is greater than we actually have. (His voice becomes quiet, vulnerable) We do so many bad things because we don't see in the long range.

Is it because we don't want to?

AN: I would say that we are not *mature* enough. We have the capacity; that could be. But so far, what have we done? (Silence) We have ruined and are still ruining so much.

You said it makes sense to assist all living beings.

AN: Every living being seems to be able to feel pain. Only humans seem to be able to clearly perceive the disagreeable situation and to help.

I have the feeling that we might not survive, but the Earth will. The Earth is so strong.

AN: There are still many, many millions of years to come for the universe. But it is not infinite. The fate of the universe is unknown. It may be that there will be a kind of chaos one thousand eighty years from now. We may joke about it, but according to some scientists, even the universe is not stable forever.

Does it matter?

AN: Yes, I feel bad about it, because the Earth has value in itself. So when it is gone, that is a terrible thing. In the long run, there will be no life possible on Earth. What can we do? It might all end in a chaos of atoms and molecules.

It is the Buddhist way of thinking: There is no beginning and no end. Everything that exists will exist in some form forever—or has maybe never existed. I have the very strong feeling that life goes on forever in some form—but that certainly doesn't make us less responsible for all we can do now and tomorrow to save the beauty around us for all living beings.

AN: Yes, I think Buddhism is the only religion I could really feel I belong to, although I don't say I am a Buddhist. But it is the best, by far! So, the success of the Buddhist way would be a great, great thing.

It is also peaceful.

AN: That is the essential part of it: the nonviolence. It seems that conflicts are inevitable in human society. But I feel strongly that conflicts, whatever their magnitude, must be solved by nonviolent means. I believe in the power of dialogue, and it is important that when we talk with people who hold different opinions from ours, our verbal communication is also nonviolent.* Some of the most important characteristics of nonviolent verbal communication are not making your opponents seem stupid and representing your opponent's views correctly.

*See Naess's book *Gandhi and Group Conflict* (Oslo: Universitetsforlaget, 1974).

Works by Arne Naess

Ecology, Community, and Lifestyle. London: Cambridge University Press, 1991.

Freedom, Emotion, and Self-Subsistence: The Structure of a Central Part of Spinoza's "Ethics." Oslo: Universitesforlaget, 1972.

Gandhi and Group Conflict. Oslo: Universitetsforlaget, 1974.

Life's Philosophy. Athens, Ga.: University of Georgia Press, 2002.

Preciseness and Interpretation. Oslo: Universitetsforlaget, 1953.

The Selected Works of Arne Naess, 11 volumes. Edited by Harold Glasser. Amsterdam: Kluwer, forthcoming.

JAMES WOLFENSOHN

The whole future of development depends on women.

James Wolfensohn and I met for the first time at a wedding lunch. I did not know the person seated next to me and asked him what interested him most in the world. His answer was short and sincere: "Everything." But what in particular? I asked him again. "Well, everything: poverty, health, housing, water management, the environment, education, women's issues . . . You must know I am the president of the World Bank."

I did not hide my embarrassment, but he did not mind and I instantly liked him, experiencing him to be a warm and genuine person. I marveled at the fact that this very man headed an institution like the World Bank. His deep concern for the world, his interest in making it a better place for all, his need to explore and to listen to as many people as possible in order to help him realize his goals (he had traveled to 120 countries since 1995)—all of this seemed proof of his care.

I wanted to pursue my discussion with him, eager to learn more about his views on the Earth. How did he see the connections between nature and economy, poverty, women's issues, and education?

On a sunny September morning in 2002,* Jessica and I walk into the World Bank in Washington, D.C., arriving for a lunch appointment with James Wolfensohn. At first glance, his influence is visible all around us: There are many women of all cultures in the building, and floods of natural light stream in from outside. Art is everywhere. The informal atmosphere is agreeable and must help people to relax, in spite of the imposing nature of the building.

At one o'clock sharp, he comes striding through the passage toward us, still talking to a man about the board meeting they had just left. Not a moment is wasted. We smile and exchange kisses, then we are ushered into a private dining room. He is relaxed and open to my questions. Wolfensohn's dark brown eyes are both sharp and warm, and he is a strong yet vulnerable man who is apparently at ease with this dichotomy.

He tells us how he grew up in a poor Jewish family in Australia. The *tsedaka*—the belief in sharing what little you have with people who need it—was part of his early education. This concept, along with music and nauture, is embedded in his life and work. He has built himself a wooden house in a forest in Wyoming to make nature his resting place and to keep his work in balance. In the seven years since he accepted the position of president of the World Bank, he feels this balance has slipped.

From time to time, he recharges in his home surrounded by wild nature, or he plays the cello. He needs these respites to cope with traveling the world for a hundred days each year, visiting projects and talking with people in the field, with community groups, and with world leaders. Attacks on the World Bank, and on himself, by various civil society groups only stimulate him to work harder, to understand more. He is acutely aware that, as the president, he has his way of thinking and his own priorities, but in an instution as large and cumbersome as is the World Bank, these won't always come through.

He is able to speak with us for two hours before he must leave for his next appointment. When we sit down around the table, I explain a bit more about the book and about the circle of people we are inter-

*At the time of our meeting in 2002, James Wolfensohn had been in office since 1995 and was in his second term as head of the World Bank.

viewing. I tell him that I have looked for a wide range of people, all of whom, in their respective fields and professions, are connected to issues related to nature and the Earth. Wolfensohn reacts with surprise and expresses that he is honored to be part of the circle that we have drawn from around the world.

We choose our food from the menus given us and begin our dialogue before the food arrives.

Given that you are often seen as a "money man," how do you feel you fit among the people in this book?

James Wolfensohn: (Reacting immediately) If I were a money man, I suppose I wouldn't be doing this job, because I think this job is much less about money than it is about equity and social values. It is much more about peace than it is about profit. The purpose of this institution is not profit; it is to bring about equitable development, which itself brings about peace. And within the framework of equitable development, as we brought out in Johannesburg [the 2002 World Summit on Sustainable Development], we can't have development unless we make it sustainable, which means caring about the world. So although it is called a bank, it is not a bank like a commercial bank. It is a fund of money, but its purpose is not profit, it is sustainable development.

In the context of that prescription, I guess I fit into a discussion about the importance of sustainability, although I don't know if my approach will be the same as that of the other people in the book.

What is nature?

JW: Well, I look at it in terms of the balance of all living things that there are on the planet, including human beings and the environment in which they live. My view of interacting is very often an interaction on a practical level—which relates essentially to poverty and sustainability of life on the planet—more than, perhaps, the spiritual level, although I will speak to the archbishop of Canterbury in October during our third meeting in a dialogue about development, because I don't think that you can look at environmental issues without thinking also about the spiritual component, which is why I started this initiative. We are bringing together the spiritual leaders. This institution is not engaged on the spiritual level, but we *are* concerned with education and development, in which those leaders can play a considerable role.

Having said that, I was just in Mongolia, and Buddhists there are assisting us on the issue of sustainability of the planet, because they have holy mountains. Buddhist leaders are approached to reach their followers at a spiritual level in relation to sustainability. It has a practical advantage to it as well, in that you preserve those holy sites. So you have a nexus between the spiritual thinking and people who don't follow that religion and don't have a religious base.

I met for a morning with the spiritual leaders because they care about damage to their mountains. What you need to do in my business is to be constantly open to crosscurrents. What I am learning is that there are many crosscurrents that you can benefit from. This is particularly true with the indigenous peoples, who live in a much closer relationship to the Earth because the Earth is part of their existence. People who grew up in Australia, like me, and who have been educated more or less in the West and have paid little attention to our own Aboriginal culture, do not know about many of these things. If you look at Americans Indians, the indigenous peoples in the Amazon, and the people in Southeast Asia, there are about two hundred and fifty million indigenous people on a planet with a total of six billion inhabitants. When you make contact, very often your only point of commonality is your contact with the Earth. Coming from Australia as I do, I understand that the whole Aborigine culture is based on their relationship with the spirits and with the land.

If I came up with this as the strategic direction of the bank, I would probably have been thrown out. On the other hand, what we are doing is absorbing all of this as we go—as an important, growing understanding of what influences sustainability.

Do you feel that you have drifted away from the old knowledge of the Aborigines, the idea of the oneness of all life?

JW: To be very honest with you, I have learned more about this in the last twenty years than when I was a kid, because growing up in Australia when I did, there was a reluctance on the part of most everybody to give value to Aboriginal culture. It is only in the last twenty years that I have started to appreciate their art and read and study that art and started to learn more about their culture. That is when I started to understand how rich and remarkable that culture is. I hope it has been retained to the benefit of Australia, but if it has been, it has been with difficulty.

So now, in this stage of your life, you are more open to the oneness of all life?

JW: I hope I have been for more than a few years. I think it has come in the last twenty years. But I also think that you are a creature of your education and environment, and mine was directed, first of all, to making a living. I have had sort of a double life. I have had a business life, but I have always spent one third of my time on nonbusiness endeavors. I started a family foundation and I have spent an enormous amount of my life on cultural and philanthropic activities. What I needed to do was to provide an adequate resource base that allowed me the luxury of expanding my horizon.

You call it a luxury. That is an interesting contradiction; the Aborigine does not need that luxury because he already has all the opportunity to spend time on nonbusiness.

JW: Yes, but you see, I am not an Aborigine and I am, therefore, both conforming to the Western framework and bound by it, to the extent of building the freedom that comes with additional resources and additional allocation of time. I can say that when I started my own business, we already put twenty percent of what we earned into social and cultural activities—from the first day. We have given away quite a lot of money, and, more important, we have participated in many activities. My kids spend a substantial part of their time on the small family foundation. They have the luxury of being able to do so, although each one has professional training and a career. But they are also, as am I, interested in those activities. My son, who started in physics and then went into music, became an accomplished composer. Two years ago, he went back to Yale to study the environment, graduated with a master's degree, and is working on environmental issues. He is deeply concerned with the balance of the planet and is making movies on the subject, devoting his life to it.

So, that is the cycle that you can have. Apart from being more clever than I am, he is able to do it because he has a lot of his practical issues taken care of. He was able to get an education in whatever subject he chose.

But then you are saying that looking after the planet or feeling the oneness of all life is necessarily a luxury.

JW: I don't doubt that there are some people who do not regard it as a luxury. But in my case, I came to it through a question of survival.

After *a question of survival.*

JW: No, I started with a question, being traditionally educated in a family without resources, and with quite a lot of financial and other challenges to face. So, I had to establish a base within that context, and I was not prepared to just forget that context, because I was educated in it. But I always had the feeling that there was something more important than making a few dollars.

I think it is important to deal with those survival questions—in my case, getting financial stability for my family and myself—and putting yourself in a position where you have choices. What I always try to do is keep a second path going in my life, which gradually has taken an increasing amount of time and effort. I was always interested in [this second path], but had to give priority to the other. But even at the very beginning of my professional career, I always gave time to nonbusiness activities. It has left me such a rich career, not financially, but in terms of being open.

That is why this job at the World Bank is so attractive to me in terms of my interests. It has humanitarian, ecological, spiritual aspects to it. Historically, the board of this institution has given little priority to cultural and spiritual issues. They are starting to gain more interest in the ethical. But when you talk about the spiritual, they will not be tremendously sympathetic to approaching the business here on the basis of the cycle of life or the contribution that individuals make, or the butterfly wings in Brazil causing a storm in Europe, or whatever the linkages are. To most civil services, that is not a big issue, any more than it would be in the Dutch civil service.

Do you think that's because people are still looking after their needs for survival?

JW: I think that most people are more interested in material security than in spiritual security.

But you were, too, in the beginning.

JW: I was, but I have always had a sense of the continuity of life and a responsibility toward others. I happen to be of the Jewish faith, and in our faith, responsibility to others is a very important element.

I grew up in a poor family, but with recognition that you are responsible for others. Whatever we had was available to others. And that made a deep impression on me, a very deep impression. So, from the early stages, whenever I had something, I always felt that I had to look after somebody else who did not have.

This is the principle of the* tsedaka*: doing something for someone whose needs are greater than your own.

JW: And you can also see it as a question of survival, because it started really as a basis for community assurance that if you did not have enough, others would try to help you.

This same spirit has prevailed historically in the United States from the time of the first settlers. Look at American society: Two hundred billion dollars a year is given to charity, which outweighs anything in other places in the world.

Yes, but it also says that in this country there are a lot of wealthy people who can afford to give some away!

JW: Sure, it does tell you that there are a lot of rich people here. But if you look at any community I've lived in, whether it's Washington, New York, or even Jackson Hole [Wyoming], and if you look at major donations, you'll find a greater number of them coming from the Jewish community. I think that's part of this sense of identifying with others and with the universe.

I came to it because I had spiritual values early on, and many people have it now. I mean, there is a return to Christianity in this country, which I think is a seeking to bring back the Christian ethic of caring, of oneness and universality. That is, I think, also effective in some forms of Islam, for example, if you look at the work Aga Khan does. He is doing fantastic work in terms of environment and humanity.

I think that many people on my board haven't caught up yet, and perhaps understandably, because they are appointed mainly from treasury backgrounds and they are concerned more with financial and development results.

You said you always knew there was something more. Your words will be recognizable to young people: I hear a lot of them say these days that after securing their lives, financially speaking, they'd like to do something useful . . . Do you think it is fear that makes people materialistic, fear of not surviving in the long run?

JW: Fear is an important element, if I consider my own background. I didn't personally want to be in a situation in which having enough money for education would be a preoccupation. But that was the case in my background.

Maybe that has a positive aspect to it as well, in that not taking our own money for granted would allow more of us to be aware of poverty in other parts of the world. I don't mean to say that all of us should live through poverty ourselves. But maybe we could understand a bit more what it means to have to face poverty.

JW: That was why I was so anxious that my kids engage themselves with people who are less fortunate than they are. It has had a profound impact on them. The one thing they will never skip is a meeting with our foundation—not only because they run it, but also because they have worked hard themselves and each has a deep concern for others.

Do you find that your attitude toward life—knowing that there is more than the materialistic side of life—is something you directly or indirectly bring to your work as director of the World Bank?

JW: This institution has changed enormously in the last seven years and I have been subject to huge criticism. But in the past year, we have done a staff survey that shows eighty-three percent of the people within the bank now agree that we are heading in the right direction.

I am doing things on the cultural level that were ridiculed in the beginning. But we are going ahead with it. Same thing with ethics and the environment. I have many, many colleagues in the bank who feel the same way I do and all they needed was to have the door opened.

When I first went to Africa, ten days after I had become president, I came back and was commenting on the trip. Someone in my audience asked me: "Now that you've visited Africa, how can you tell what constitutes a good project?" I answered that I could tell by the smile on a child's face. I was ridiculed by a lot of people for the answer, although there were a few who thought: "Maybe that human aspect *is* what really counts." And nowadays, many more people might agree that you can look at the smiles to check how the projects are going.

Not only have people within the bank been skeptics. Through the years both the institution and Wolfensohn have been under constant fire:

I would like to talk about the image of the World Bank. The institution is often severely criticized.

JW: Well, it depends where.

In Johannesburg, for example, at the World Summit on Sustainable Development.

I show Wolfensohn an angry political ad from the Herald Tribune *in which he, George W. Bush, and the chairman of Citicorp Financial Group, Sandy Weill, are aligned as three who do not care about equity and who support the evils of globalization.*

JW: But it [the ad] did have absolutely no impact, I am happy to say. On the contrary, I had forty people there and we have never been more successful than in Johannesburg. We were in every meeting, participated in every dialogue. The press was very kind to us and I think, in many ways, we were the intellectual leader in Johannesburg.

There will always be criticism too, and there should be. There should always be a critical civil society, saying, "Do this or do that." I think many of the things the bank is doing now are things we would not have done if it had not been for advanced leadership in civil society. But on the major issues today, we have reached a very interesting equilibrium, because the framework for development was very broadly agreed on in Johannesburg, and that framework sets responsibilities for the rich and the poor countries. It states that if there is to be equitable and sustainable development, no one can do it alone—not civil society, not the bank, nobody can act alone. There is a need now for partnership, which should include, under the leadership of the governments of developing countries, multilateral and bilateral institutions, the private sector, and civil society—because if we all work separately, we will never meet the goals. The leaders of civil society do really care about the issues, and I understand this extremely well, and we reach out.

This new form of realism is built on the understanding that if you don't create partnerships, issues like the environment, AIDS, education, health, water, energy (all of which are built around the question of poverty) are not going anywhere. Civil society cannot be fully effective without the World Bank, and the World Bank is not going to do it without civil society and the private sector. There is a growing respect—which is different from agreement—on every subject.

Listening to his words, we can recognize that many of the practices Wolfensohn brought into the bank must have been new to a lot of the people working there. This must make him feel loneliness at times—the loneliness that comes with being a pioneer making changes. I am curious how he sees that role within a large, cumbersome institution.

Do you consider yourself a pioneer?

JW: I think I am building on what people have done before, making some important connections in terms of the unity of nations, in terms of the breadth of things that have to be dealt with in the field of development. I started an approach called *comprehensive development.* Basically, I say that we are not going to get anywhere if you don't look at development comprehensively. I started with the need for legal and judicial and financial sector reforms. Fighting corruption is something without which you will not get things done. Then I looked into education and health, followed by issues of environment and culture. Within culture, I started to talk about spiritual values.

I am convinced that without all this, you will not have sustainable development. And if you don't have that sustainability, it will be no more than just another World Bank project being imposed on people. It has to come from within each country.

At this time in our conversation, Wolfensohn offers us dessert—ice cream with chocolate sauce and fruit—but we prefer not to interrupt our conversation. He continues explaining his vision of comprehensive development, so needed, in his eyes, to really make progress.

JW: When I was in Central America, I met with Mayan leaders, and we had just built a new school. The school literally was a redbrick schoolhouse. It could have been in Connecticut. It was a very modern design. Next door stood a typical Mayan house of assembly in which I met with the old people [who lived in that community]. They were incredibly proud of their tradition and knowledge of nature and astronomy, which goes back a millennium.

I came back here to Washington and said: You people are crazy. You are putting a modern school in an area with a Mayan tradition. It would cost nothing to make that school look Mayan. I don't mean to make changes to the layout or the floor plan, but to somehow engage the people there in the construction. You must create institutions grounded in their beliefs, and they have to have links to their past.

The same thing happened in Mali, with a centuries-old tradition of greatness extending all the way back to ancient Egypt. If we are going to build there, we have to build on their history, culture, and beliefs. Since then, we have tried to build upon the Malian culture so that many people today do identify with their tradition [in the new construction].

Tell us more about your strong desire to promote women—especially indigenous women—in the organization.

JW: I have a managing director—one of four—who is South African. Mamphela Ramphele had a significant role in the struggle for independence [in her country] and was subsequently appointed vice chancellor of Cape Town University. She is now one of our managing directors.

Why do you choose women?

JW: I am choosing talent, not women.

Is there an extra value that you think women bring?

JW: When I arrived in the job, we didn't have one woman working on social issues such as education and health. (With a twinkle in his eye) And since women, as you might have noticed, *are* different from men . . . I have worked on the board of the Population Council for almost thirty years. The thing that upset me most was that in an institution like this bank, which is talking about gender equality, there seemed to me to be too little equality on our own work floor.

But what I'm doing I don't do for some kind of justice. I do it because the place becomes much richer. If you have a bunch of white men coming together to make decisions, you will get white men's decisions. If you have a diverse body of men and women, you at least have a better reflection of society. And if you have what we call here part one and part two—namely rich and poor—to make decisions, you are more effectively approximating the world.

Also, when I came here, you couldn't assign someone who came from a certain country to work in that country. They thought he or she would be biased. I changed that, and the person running the African department for the last four years is actually an African. He does a magnificent job and knows more about Africa than I will ever know—more than any non-African person will ever know about his continent.

Working with different cultures, and working with both women and men, I suppose, makes it possible for you to learn and to broaden your vision every day.

JW: Of course. If you go to the cafeteria, you'll see that we have ten different kitchens. It is truly a global institution.

Yes, but I am also referring to different ways of thinking, different ways of solving problems. I suppose the same word will be received in a different way by you than by, for instance, your African manager.

JW: Oh yes, it is! In an institution like this, we want to synthesize lessons that you learn from different cultures—and there are a number of common experiences. When you come down to working with individual cultures, as I said earlier, you have to build your intervention on something that is culturally appropriate. So you may have beliefs in relation to carbon emission and water environment, issues that are global in scope and which apply generally equally around the world. But these might have very different impacts depending on the culture in which you are operating. Water in the Amazon has a different meaning from what it has in the Sahara. It is used for different reasons.

In the work that we are doing here, I proposed a second big issue in the comprehensive approach to development: The methodology for establishing a program in a country had to provide time and opportunity to ask the people and the government of that country for leadership and counsel. So now we have programs that are established by governments, but only after consultation with civil society and the citizens in that country. That means that a Malian may be asked to give views on issues affecting a rural area in his or her country before action is taken. What we are seeking to do is to broaden the range of consultation beyond whoever is in power, beyond the Malian elite, in this case. If you think there is a gap between white Europeans and the people in the field, what I discovered is that between the many elites in Africa, there is a gap at least as large. And very frequently, there's far less willingness to bridge that gap among the privileged of the local society. So when I visit an African country, my typical routine is that of the four days of my stay, I spend two and a half days in the field and a day and a half in the capital. I am nearly always in the field before I sit down with the president of a country. My stay in the field is the best part of my trip, talking to people in rural communities and urban slum areas.

Going back to the issue of women: Do you see a difference, now that you have brought so many more women into this institution? What has changed with more women on board?

JW: I think this place is changing very much, but too slowly. What I can't . . . (Pausing to reflect) If a man is doing a terrific job, I can't fire him just to make my gender equality better. If an opportunity comes up, I have to be as open as possible to give both men and women a fair chance.

But we have targets. We have someone here whose only concern is gender equality. And I have just taken on an adviser on disabilities. Because the other area that is globally marginalized, apart from that of the indigenous peoples, deals with disabilities. In every society, you have between five and ten percent of the people with some form of disability—in *every* society. And in developing countries, it is greater, usually including disabilities caused by war and malnutrition and not including mental disabilities, which is a huge number in itself. So, if you are looking at society in general, you have to think in terms of disabilities, too. I have worked in the disabled movement—I ran the global Multiple Sclerosis Society—so I have been very conscious of that.

Returning to women once more: In all cultures, women used to tend the land. Then technology arose, mostly dominated by men, and the women disappeared. Is this your understanding as well?

JW: I would disagree with that. I think that today, the dominant factor in developing countries is still women. They are not recognized, but they do the work and hold the families together. The whole future of development depends on women. I am not saying this here because both of you are women, but because I say it publicly many times. It's true. Therefore, it is important that we give opportunities to women, whether it is in microcredits or profit rights or even protection against violence. We have done a study—I don't know whether you have seen it—named "Voices of the Poor." We interviewed sixty thousand poor people and came out with a report. What fascinated me was what the poor wanted most in life: Very rarely did they talk about money. They *did* talk about recognition, rights, a voice. But by far the most important thing for women was physical protection. The issue of abuse of women is a global theme.

Probably mostly because of alcoholism?

JW: Partly that, but also out of jealousy. And sometimes it has to do with cultural history, or it is even religiously related.

The principal keys to development, I would say, are health care and property rights for women, because in most cases women don't have property rights. That is a huge issue, in my view.

Would you add education? You have written that by 2015, all children around the world should have access to education.

JW: Absolutely, and I did not say it inadvertently. Education is central. That's what the Millennium Development Goals espouse. That's the thing that has been agreed to by the United Nations. At the annual meeting of the United Nations in a couple of weeks I'll be giving a speech again on this theme that I touched on in Johannesburg. It will be about what we all have to do.

I recently had some long talks with [Thabo] Mbeki, when he was here in Washington. He is very clear on what Africa should do. I said to him that the fact of the matter is that there is huge agreement now on what the developed world should do in terms of traditional aid and access to trade. And there is agreement on the part of developing countries' leaders that they should strengthen their capacity to fight corruption and develop legal systems, and that they must work to implement the Millennium Development Goals, which focus on education, reduced infant mortality, and environmental issues.

I think what happened in Johannesburg is that we reached the point where there is no more denying what is needed. Maybe it didn't add much to the discussion, but the good thing about the conference is that it endorsed what people have been saying. So for me—and that will be the theme in my speech at the United Nations—there is nowhere to hide! Everybody—I have selected, for example, some statements from President Bush, Chirac, and Schröder—has said what he is going to do. Now, the question is: Will they do it? That is where we are. We have to implement these policies, we have to deal with the questions that everyone agreed on, including gender equality.

And there is the problem of AIDS. Mbeki has not handled the issue very well.

JW: That's his weakest point. But he's actually stepping up now.

As president of the World Bank, you have stressed more than once the importance of education in changing the position of developing countries. You actually expressed it quite beautifully: Education should teach toward love, not toward hate.

JW: You see, if you have people teaching hate, you will have a generation that will be brought up by hate. Unfortunately, many extremists, religious and political, teach hate to young children. If that continues, peace is almost impossible. And I am very critical of that.

Your term as president of the bank ends in 2005. What is your dream for the bank in today's world?

JW: I think the World Bank is making and can make a unique contribution to bringing about social justice. We can do that in a number of ways: certainly, by addressing social issues and by financing. But I feel our role is much more important than that, because whether we lend a few dollars more or less is not making the difference. What *is* going to make a difference is coordinating ideas and helping to build partnerships. The bank's contribution is to make the world understand that development is not something that happens from day to day or from year to year. The *one* thing the bank can do is provide continuity over time. To be successful, most of these programs will have to become self-sustaining and last for years. What I hope the bank will do is retain knowledge and focus in a way that will carry through programs initiated today or those that have been initiated over a period of time, so that the full effect of good thoughts today can be felt twenty years from now.

You can get everybody to agree now on programs for education for all. We would all feel terrific if we succeeded in helping a few extra kids get to school by starting such programs. But quite frankly, unless it is carried on for at least ten years, it is not going to be embedded in a society. You have to go through a cycle of education, which includes seeing the girls and boys coming out of school and finding jobs, then find that their families adjust and that the family size has changed because the children are becoming productive, and, finally, that the families have opportunities to move into new areas. That is full cycle to me. Just educating children is not going to be sufficient. So education is the key, but it is not the only issue. You must have economic growth to create jobs. You have to educate people and create an environment in which people with education can, in fact, change and be productive.

As I said before, I think development is a complicated, comprehensive thing. And it is fine to start with a priority on education, or on health care, or on carbon emissions, or whatever. But we shouldn't kid ourselves that we are going to achieve instant success on any of these issues. One of the principal roles of the bank is to provide an ongoing measurement and context, and we do that in many ways as the governments change.

Take Mexico: When the government in Mexico changed two years ago, we gave them a four-hundred-page book in which we pointed out issues the government needed to face. It was meant as a bridge, a helping hand, to the new government for them to consider when establishing their own programs. When Brazil's government changes by the end of the year, we will have already looked at everything that has been done in Brazil.

We review the countries every year, but it becomes particularly important to provide them with the full picture when they have a governmental change. When Vladimir Putin came to power, I had known him beforehand and I said to him, four weeks before the elections: "The best thing I could do for you would be to bring in a dozen of my colleagues to give you a straight view of what we think are the issues confronting Russia. Then *you* can formulate *your* ideas on where you want to go with the country." I told him that we preferred to do it outside the Kremlin, and we should have a man-to-man discussion. He said, "Would you come?" I said yes. He asked, "How about Sunday a week?" I answered, "But that's one week before the elections!" And then he said, "I am not going to work on Sunday, so I will arrange it."

I talked to ten or twelve of my colleagues and we sat from eight in the morning till two in the afternoon with Putin. We discussed everything, and that may have helped him prepare for his first speech to the country, which talked about more democratic and more market-oriented values. I know that that day had a huge impact on what he said. We have become good friends since, so I see him a lot. I respect him enormously. He's a great leader.

And how does Putin, in this case, know that you are not talking from an American point of view but instead are there to provide him with more neutral information?

JW: Because of the way we were talking. My colleagues are independent.

How many days a year are you abroad?

JW: Probably about ninety to a hundred. I have been to approximately a hundred and twenty countries since I became president of the bank.

At this point, he offers us a second cup of tea. He orders a decaf for himself. I seize the opportunity to introduce the subject of September 11. It feels impossible not to talk about it and its impact on the bank's future and, in light of the subject of this book, on the Earth.

JW: I think the deep effect of September 11 is that [before it] people existed with the fiction that there were two worlds: There was the developed world, the rich world, and there was the developing world—one world with six billion people in which we find eighty percent of the world population holding less than twenty percent of its wealth and the other ten percent of the world dividing the rest of the wealth.

I think most people were living with the fiction that these two worlds were separate. I'm sure you didn't think it, and I didn't believe it, and many others didn't. We have seen that whatever you think of globalization or however you describe it, the world has become smaller through trade, finance, environment, health, drugs, terror, immigration, and so on. Maybe people like us already understood that. But I think when Afghanistan and Saudi Arabia came to the World Trade Center on Wall Street, to Pennsylvania, and to the Pentagon—for me the impact of September 11 is that the wall came down. I often use the collapse of the World Trade Center as symbolic of the collapse of that imaginary wall. No longer could we believe that problems in distant lands could not affect us here. There was an enormous change of heart in this country, and a widespread recognition of a single, integrated world.

But it didn't help, because the doors have been shut even more.

JW: That is the current leadership, but not the people in this country—I mean, I can tell when I go to speak. Where there were two hundred people, I get five hundred now. I am not saying that egocentrically or arrogantly, but I think the interest has grown. People want to know.

The language has changed in this country: You can go out to the Midwest and fill a room with people wanting to talk about international, global issues. People want to hear now, and that was much less common before.

Is television adding anything to that?

JW: Oh yes, I think it does. The difference is that before, you would look at the war in Somalia and think that it was miles away from home. Today, you think: If something is happening in Somalia now, could that have a positive or negative impact in this country?

After talking about major world issues, I still have some personal questions to ask Wolfensohn, so I radically change the subject, assuming that he is accustomed to such shifts:

Have you ever experienced the oneness of nature? Have you lived moments in which you have felt that oneness, being part of . . .

JW: (Exclaiming) Oh yes, lots! I used to go to Alaska every year. I went for eighteen years in a row, fishing for salmon in the river system. A bear would come somewhere down the stream, and I felt the fear of the bear while watching nature at the same time. That is my strongest image of nature. I stood there in the river, with the water high up against my thighs, and there suddenly is this brown bear approaching. That was a moment when I felt one with nature.

Also, when I can't sleep, I think about the river in front of my house in Jackson Hole, Wyoming. I have a house there, a wooden house built into the natural surrounding. That is why I built it. I can feel the quietness there, which you can't feel just anywhere. (Laughing) Certainly not on Fifth Avenue. But you can feel that in Jackson Hole, Wyoming, which is why I want to retire there.

Rupert Sheldrake talks about those in England who spend their weekdays making decisions that go against the needs of nature, or living removed from nature, and then go to their houses in the country on weekends to surround themselves with nature and revitalize their bodies and souls. In a way, we are all more or less guilty of this, aren't we?

JW: You see, what I decided twelve, fourteen years ago is to make my center *in* nature and then reverse what you just described. It started off very well with my own business, until seven years ago, when I took this job, and now it has flipped on its edge . . . The other thing that is spiritual to me is music. In music I feel a different dimension. I picked up the cello when I was forty-one, and now playing music and chamber music take me to this other dimension. Nature, in a sense, has that same power.

Among all his other accomplishments, Wolfensohn has been chairman of Carnegie Hall in New York and once organized a special birthday party there, which included his giving a concert with some famous musical friends. As we talk, we learn that he is planning another concert there for his seventieth birthday, in cooperation with well-known musicians such as Yo-Yo Ma.

JW: We are thinking of doing the Mendelssohn Octet with sixteen young artists. We want to invite Arab kids, Israeli kids, and children from Pakistan, India, Ethiopia, and Eritrea to play. It originally is, of course, for only eight instruments, but it can be done with a chamber orchestra. I guess the kids will love to play with great musicians like Yo-Yo Ma and my friend Pinchas Zuckerman. Daniel Barenboim does this work every year with young people from the Arab-Israeli youth.

I'm in nature when I'm literally [outside], in nature itself, and through music in the sense that both make me feel I'm in another world, in the world of the spiritual.

Most of the other people whom we have interviewed are working more directly with nature. I have talked with all of them about the consciousness that is in every life-form. Can you imagine that there is awareness in flowers, trees?

JW: (Measured silence) . . . That is a stretch for me! But I have a great respect for biodiversity, for instance. We started a major initiative with Conservation International to preserve biodiversity because of the interaction between the Earth's diversity—between the species, flowers and so forth, caused by their urge to survive—and the balance of nature. That much I do understand and believe in.

Could you see awareness in, say, a lion?

JW: I am conscious . . . I can reach the animal kingdom in terms of awareness. But if you talk about awareness in flowers or physical objects, I guess I haven't got so far, as yet, to understand their sensitivities.

But I do respect people who do feel that. If you see my house, it is surrounded by wildflowers and stones, because it is the balance of nature that we need. So I do appreciate it and understand the continuity.

There is an energy that every life-form carries inside, like a light that animates it, and this is received by other life-forms that approach it, whether deer or birds or other living things. There is a continuous interchange or conversation going on between all these energies that communicates their individual essences . . .

JW: I could be convinced of that, but I haven't studied it enough. Sorry to let you down . . .

On the contrary: There are many people who think about it in the same way that you do, who can relate to your way of looking at these themes. Thinking about them will help people realize at least that there is life in all that surrounds us. And this might change attitudes . . . How do you think the World Bank can stimulate the relationship between humans and nature?

JW: The first thing we can do is understand the best way to preserve nature and provide the right balance. We have the Global Environment Facility engaged in environmental analysis that will allow for the preservation of nature—and for a balance of nature with economic development, which is not always easy. In fact, we very often are severely criticized for the choices we make regarding it.

A very recent example is the Chad-Cameroon pipeline. The discussion was over whether we need such a pipeline, because it would have a serious impact on nature. We made the deal, trying to get that four billion dollars of income to Chad. You see, oil is its only asset, providing its greatest source of income. Our responsibility was to build an environmentally sensitive pipeline, which we have done. I'm very proud that it's up and running. Now, my concern is no longer environmental, but it is to ensure that money goes to developmental and social purposes. But for some of the environmental groups, it was unethical to build the pipeline. I believe that you have to make choices at times, trying to work with nature as much as possible. The same is true for some dams and water projects, considering the growing world population. We'll reach eight billion people on this planet in the next twenty-five years. We'll be forced to deal with the questions of famine, water, sanitation, education—all those important issues. And for that, we'll need money and physical resources. The balance in those terms is very difficult. That is why I consider it helpful to have NGOs [nongovernmental organizations, usually nonprofits] balancing what you are doing. There will be more dilemmas to come.

Given that you are in favor of comprehensive development, can you tell us an example of how the World Bank has changed its policy for financing projects since you have been in office?

JW: In nearly all of our projects today, we look at the interaction of each aspect of the development process, from the strength of governance to the health of the community to the environment, and we seek for civil society and the private sector to participate in design and implementation.

Can you give an example of a project that has been criticized in the past and which you would no longer finance today?

JW: Probably one of our early dam projects or road projects. But today we would do a better, more sensitive design. The developing world still needs water and communications.

Just as I'm ready to ask Wolfensohn my last question, an aide walks into the room to let him know that the finance minister of Japan has arrived.

What is love?

JW: (After a long silence, then taking his time) Probably, the most precious thing of all. It is the purest form of association between individuals, and maybe between an individual and the world around him. Also, it is the strongest tie that you can have, because it is a tie that goes beyond the physical or intellectual to the essence of humanity.

The war with Iraq began after we had this conversation with James Wolfensohn. With this significant development, I felt the need to contact him again and ask if he finds continued interest, or still-growing interest, in his lectures on international issues in the United States.

Do you find that people still want to hear about global issues?

JW: I doubt that there is anyone today who is not interested or searching for answers about the new balance that will one day come to the world. My hope is that it will be more equitable and that it will be peaceful. I know of no one who does not hope for peace for future generations. My colleagues and I at the bank will continue to work for this goal.

Works by James Wolfensohn

A Case for Aid: Building Consensus for Development Assistance. Washington, D.C.: World Bank, 2002.

PATRICIA MISCHE

Peace with the earth is essential
for peace with one another.

Flying from Washington, D.C., to Dayton, Ohio, we stop first in Chicago, where we board a small plane to bring us south again to our final destination. The plane is crowded and the ride bumpy, and we are not too happy, but we make it to Dayton and all is well when Patricia meets us at the airport.

I met Patricia for the first time at a peace conference in the Netherlands, where we both participated in a forum on education. It was then that we connected, realizing our passions lay in the same direction. On that occasion she dumped on my lap a huge pile of documents about Global Education Associates (GEA), an organization she founded with her husband in 1973. It consists of a network of men and women in ninety countries who are engaged in research, education, and action to advance global systems that will secure ecological integrity, peace, social justice, and democratic participation for present and future generations.

For three years after her husband's death, she continued to lead the organization as its president, chief administrator, fund-raiser, editor, and speaker, but in 1999, Patricia accepted an invitation to become Lloyd Professor of Peace Studies and World Law at Antioch University in Yellow Springs, Ohio. She remained on the board of GEA and recently, while continuing to teach, resumed leadership as chair of the board.

Patricia has an exceptional, clear mind and, with it, an original, free way of thinking. It is wonderful to see her again. On the way to her house, she explains how Antioch University embodies a freedom of spirit that she enjoys immensely.

Her wooden house is comfortable. For dinner, she picks us some fresh vegetables and herbs from her garden. Pictures of her grandchildren are fastened to the fridge and papers and books dominate the rooms.

Patricia was among the first people who saw the need for a covenant with the Earth. In 1988, at a citizen diplomacy meeting between U.S. and Soviet citizens, she presented the idea for a citizens treaty on global ecological security. Following up on this, she coordinated an international effort involving people in one hundred countries to gather input, write, and promote such a treaty. Completed in 1989, the Earth Covenant was signed by more than two million people by the time of the 1992 Earth Summit in Rio de Janeiro. It was endorsed by the United Nations Environmental Program (UNEP) and served as a prototype for more than forty citizen treaties developed at the Nongovernmental Organization (NGO) Forum in Rio and, later, for the now well-known Earth Charter.

After dinner, we settle on cozy couches in a spacious room. I am longing to start our conversation and to listen to what Patricia has to say in response to my first question.

What is nature?

Patricia Mische: (Taking her time) It is a community of life. I guess that is my best definition.

What is your connection with that community of life?

PM: My earliest connections were when I was very, very young. I have always felt very connected to nature, and I don't quite know why—except when I think of my father and mother. I was born at the beginning of World War II and my family kept a garden for survival. Or better, my mother kept the garden; my father was away in the Navy during the war.

Out of this garden came everything we ate. It was really survival. We were relatively poor at the time, and for some years everything came from that garden. We grew potatoes, lettuce, carrots, tomatoes . . . everything.

My father had wanted to be a forest ranger. He was in love with nature. But his father died when he was still at school, and as the oldest son in a family with thirteen children, he left school and went to work in the very foundry that had killed his father. His father had died of silicosis, with lungs as hard as concrete from breathing foundry pollutants. My father worked in factories all his life, his hopes of a life working among trees forever dashed. But he never lost his love of nature. He would come home from the assembly line and walk straight to the family garden. He had planted flowers, and would stand in the garden for a while, looking at them and the vegetables before he came into the house. It was a way to heal himself from the noise and hard monotony of the factory, a way to reconstitute his soul and rehumanize himself after what the factory had done to him that day, before joining his wife and six children.

Despite the harsh conditions of his life, my father loved nature and found ways to stay close to nature till the end of his days. And from an early age, I remember feeling a great attraction to nature, too. That is where I knew God, where I would find meaning, inspiration, and focus in my life. It was very important to me.

We lived at the edge of a small town next to farms and cornfields. When I needed to think, I would walk into the fields or ride my bicycle on backcountry roads. For recreation, my parents went fishing in nearby lakes. I went with them, but not to fish; I just enjoyed being in the vibrant community of life in and around the lakes. Just being there, aware of an incredible richness of life, was what mattered. I had a sense of being part of something vast and meaningful. I felt this great intimacy, closeness to other beings and to life. I had a true sense of unity there and a delight in seeing everything growing, alive, and interacting.

In your father's footsteps! How does the connection with nature you had when you were a child influence your life and work now?

PM: My best spiritual inspiration and healing comes when I am in the presence of natural beauty and feel part of a community of life, whether this is in mountains, by oceans, or with trees in a forest. I have lived in many different parts of the world, including in cities; I lived in New York for quite a while, and I loved it. So I don't have anything against cities. But what "saved" me in New York was Central Park and Riverside Park. Good park systems within an urban area are oases where human life and the life of nature come together. In Central Park, people were replanting and restoring the species of plants that once grew there. This in turn attracted certain birds and animals to return that had not been seen there for a hundred years. For example, once the marsh grasses they favor were replanted in the north end of the park, snow egrets returned and now mingle with the people!

I found that there are ways to live in a highly developed and busy city and still experience nature. But I often get so busy with daily demands that I forget to take time to do this.

You are saying that you are not bringing nature into your working life as much as you would like to. But at the same time, you teach about peace in a way that is very connected to all life.

PM: For thirty years I have been reading and thinking about the need and content of a new cosmology and Earth consciousness. Rather than trying to deal with environmental issues just as problems, we need to develop a full vision and deepened consciousness of Earth as a living organism or community of life, and of its evolution in time and our place in it. As part of my work with GEA, I was editing *The Whole Earth Papers* and *Breakthrough* and, through these vehicles, trying to bring together the best thinking on various aspects of a new cosmology. I try to do the same thing in my teaching, but in the process I often allow myself to become too busy with details of nurturing and imparting to others and forget to work on developing my own spirit. Academically, I may be growing. I am doing my work. But I don't take enough time for my own soul work.

If you hadn't felt so close to nature as a child, you wouldn't have brought it into your intellectual work.

PM: That's true. And my earliest spirituality also evolved through relationships with nature.

My husband and I wrote a book that got good reviews for its political analysis and also for its focus on values and spirit. But one conservative theologian criticized the underlying spirituality as pantheistic. Another who heard me speak on themes in the book cast me in with the anti-Christ!

It didn't bother you?

PM: (Laughing) I thought it was funny. He ascribed far more power to me than I actually had.

Talking about both her intellectual and spiritual career, I want to know more about the organization, Global Education Associates (GEA), that Patricia and her husband, Gerald, founded in the seventies.

PM: We founded GEA in 1973. It grew out of our international experience and a sense of how global forces were going to shape a different world. Before we married, my husband, Jerry, had lived and worked in Mexico for a while and then founded the Association for International Development (AID), in which primarily North and South Americans worked together to train leaders for economic and community development. And I had studied and taught in Uganda and Kenya for several years as part of Teachers for East Africa. I was there as these countries became independent from Britain and struggled to form new nation-states.

Also while I was in East Africa, John Glenn was orbiting the Earth and satellite-monitoring stations had been set up along the equator, near where I was teaching. Meanwhile, the Leakeys had discovered the oldest human fossils not far from where I was teaching. My students came from eight different tribal groups and in only a few years had to move from tribal to new national identities and thence into an interdependent world and an emerging space age.

Incredible historic transformations were occurring all around me. My sense of the human past and human future opened to hitherto unknown vistas. It was soon clear that nation-states were a totally inadequate way in which to view the world and that all states would soon face challenges from a rapidly globalizing world. The nation-state system would be an inadequate framework for addressing new global forces. We started Global Education Associates to promote international and multicultural dialogue about the vision, spirit, and systems that would be needed to effectively address our common global future.

While I was in Africa, I immersed myself in African culture and learned to see the world through African eyes. This was very important in my development and future work. To be with people who were very close to nature—and who were very close to losing that contact through modernization processes—was very formative. I fell in love with the African people and their cultures.

And the Earth?

PM: Well, I had loved the Earth before. But I came to love the African Earth, yes. It was such a special time.

It's funny how the Leakeys come up in so many stories! For example, Jane Goodall talked about them: Robert Leakey was the one who encouraged her to go to university to prove to the scientific world what she already knew. He foresaw that otherwise she would never be accepted.

PM: They were working in Kenya and Tanzania when I was there. Richard Leakey—the son of Mary and Louis—was born and raised there. I never actually met the Leakeys, but my husband met Richard. He was an adviser to Global Education Associates when we first started the organization.

Your husband worked in Mexico and you were in Kenya and at that time you didn't know each other yet?

PM: I met Jerry after he had come back from Mexico and founded AID. As a college student, I became interested in international service and organized events that brought a variety of people to campus who were working internationally. Jerry was one of the presenters. We got to know each other better when I was preparing to go to Africa, and then, during my several years there, we corresponded regularly and were married when I returned. We worked in different world regions and later were able to pool our experience and contacts in the work of Global Education Associates. In our development work, we had seen how some people were helped—there were many success stories. But if we looked at the macroeconomic picture, we saw a different story. Decisions made in other parts of the world by powerful economic entities could overturn these small successes in a day. That's why we felt the need to start an organization to look at the impact of existing global systems on local realities, and to explore just and equitable alternatives.

From the beginning, we also included environmental concerns as part of our focus. Rachel Carson's book *Silent Spring* had made an impact around the world. She exposed the effects that pesticides and other pollutants were having on birds and the life system. She saw the death of birds as an early warning of what was happening to the whole life system and what could happen to human health if we did not change our behavior and relationship with nature.

That was before the think tank the Club of Rome* came to be?

PM: Yes. The book helped catalyze environmental movements in Japan, the United States, and Europe. It was translated into many languages and read all over the world. It led to action in the United Nations. Carson was a scientist, but some vested interests tried to undermine her work by ridiculing her as a hysterical woman. She persisted, however, and her work inspired the modern environmental movement. Al Gore said that his environmentalism grew from reading Carson's work. Later,

---

*As a nonprofit, nongovernmental organization (NGO), the Club of Rome "brings together scientists, economists, businessmen, international high civil servants, heads of state and former heads of state from all five continents who are convinced that the future of humankind is not determined once and for all and that each human being can contribute to the improvement of our societies." See www.clubofrome.org.

as the vice president of the United States—more than twenty years after the book was published—Gore lamented that "since the publication of *Silent Spring,* the legal, regulatory, and political system has failed to respond adequately."*

Ecology as a science began to evolve and spread. In the 1960s in Japan, the environmental movement started to grow after people died or suffered serious health problems from industrial pollution. In the United States, the movements started to become more visible, too. It was not long before the findings of ecologists and other scientists and the work of environmental movements began to have an impact on international development programs and UN agencies. Some of the early development programs promoted or funded by the World Bank, IMF [International Monetary Fund], UNDP [United Nations Development Program], FAO [Food and Agriculture Organization], and other UN-related agencies had been nonsustainable or even harmful. Traditional practices and wisdom had been ignored, with negative consequences. While I was in Africa, big companies were buying up the land and using it to produce export crops or national governments were pressing people to plant sugarcane, tobacco, tea, or coffee instead of the usual foods that fed their families. They were lured by the promise of more income. But in a very short time, where once people had been food self-sufficient, they were now hungry! Tobacco and sugarcane did not provide nutrition for their children, and, soon, not even cash to buy food. Monocropping wore out the land. Commodity crops such as sugarcane soon glutted the local and international markets and the prices local planters received dropped below their production and replanting costs. People who had been food-independent now needed "care packages" to survive. Thus, in 1960s Africa, we could already see the early stages of what today is called globalization.

Later, after I was back in the United States, I continued reading the work of Teilhard de Chardin. Teilhard, along with his Russian contemporary Vladimir Vernadsky, through his studies of evolution, paleontology, geology, and cosmology, had arrived at the conclusion that

---

*Source: www.students.haverford.edu/wmbweb/medbios/bbcarson.html.

human consciousness was increasingly important in shaping the further stages of planetary evolution. Humans had emerged out of cosmological and planetary processes. They were from the Earth and of the Earth. Their bodies comprised the same material elements of the Earth in the same proportion that existed in the Earth. However, humans were the Earth in a new, macro-phase of its evolution—the consciousness phase. They asserted that with the emergence of human consciousness, a certain threshold in evolution had been crossed beyond which the further stages of evolution—indeed, the fate of the Earth—would increasingly be determined by human consciousness. Human beings initially quietly entered the Earth and its evolutionary processes, learning to survive amid the more powerful forces of nature and find a niche within Earth's life systems. Their penchant for learning gave them increasing powers, and in the last fifty thousand years, the human species had spread all around the planet and become a total and dominating presence. This hominization of the Earth was the first "globalization."

By the 1960s and early 1970s, evidence was amassing that increasing and global-scale harm to Earth's community of life was resulting from human activities. The problem was not just at the surface level of visible human activities. The harm grew from the lack of modern consciousness of humans' integral relationship with other beings in the community of life. If the Earth and humans, whose lives and health depend on the Earth's healthy functioning, were to be healed, human consciousness would have to change and deepen. The modern paradigm of human domination over the Earth had to give way to a profound, living awareness of the community of life, and we had to learn to live as responsible members of this larger community. We needed to see how degradation of this life community affects our own, human well-being physically, mentally, and spiritually.

Those were some things that began to come together for me in the sixties and seventies. Earlier in my life, I found my fulfillment in feeling part of nature. But now I became aware of the pain and damage that was being done to this life community and the necessity for each of us to try to make a difference, to be more aware, more conscious of how we think and what we do. I suppose my sense of connection at this time included great sadness, but also a sense of great responsibility to find an adequate response to what I was seeing. As an educator, it seemed my best response would be to facilitate new learning. I felt that there was a great bottleneck in the educational process. Teachers themselves lacked the necessary awareness and training for teaching others. To help teachers prepare to teach others would be much more than teaching one classroom of students. It would have a multiplier effect, as each teacher taught hundreds of others. So, in 1973, I started the first teacher-education program in the United States related to new consciousness and education for peace, justice, and environmental concerns. This was also the year we founded GEA, and the institute for teachers was our first annual program. Jerry and I also began work on a book that came out a few years later under the title *Toward a Human World Order,* and we began publishing a series of papers called *The Whole Earth Papers,* which explored new consciousness and strategies for transformation. We were soon being invited to speak and to run workshops in India, Japan, the Philippines, Africa, Latin America, and Europe, as well as all over North America.

But the busier I got doing all this, the greater was the risk of becoming disconnected from the very source of my vision and inspiration. I did not have time to be in nature for my own renewal. After a while, you dry up.

You have said, "Peace on Earth cannot be realized without peace with Earth." That is quite a statement. We are not lovers of the Earth anymore.

PM: We have become so alienated from the larger life community that we forget our integral relationship to other beings in this community. We are alienated from our own true self. (Laughing ruefully, almost apologetically) We don't know why we are unhappy.

We are not at peace with ourselves because we are not at peace with the larger life community. We don't belong anymore; we don't have a sense of place. We are moving in an alienating and alienated state within the community but not even recognizing the community's existence.

The degradation of the Earth really does lead to war. There is some evidence that the system of organized warfare, which did not exist among our earliest human ancestors, but emerged only in the last ten thousand years, arose from environmental degradation. The degradation may have been caused by natural disasters, or more likely such human activities as overgrazing, overhunting, and other unsustainable practices. When degraded ecosystems could no longer support the needs of growing human populations, some groups moved to new territories. But over time, some dealt with scarcities of food, water, or other basic essentials by raiding and plundering neighboring or even distant communities. At first these raids may have been intermittent and not a dominant ethos. But over time, as raiding paid off in greater resource security, and as warriors were bestowed by security-conscious populations with enhanced social status and power, whole social systems and belief systems began to be organized around warfare and preparedness to defend against real or imagined threats to security. From tribal groups competing for land, food, water, and other vital resources to modern nation-states competing for oil, balance of payments, markets, and more, the war system grew from periodic raids to a full-fledged war system with threats of mass destruction.

In the future, as increasing populations compete for diminishing resources, we can expect even more conflict, if humans do not change their consciousness and patterns of relationship with the Earth and one another. Peace with the Earth is essential for peace with one another. At the same time, peace with one another is essential to peace with the Earth. War and war preparations with new technologies of mass destruction are now so damaging to the Earth that regardless of whatever utility war may once have had in ensuring group security, it now undermines the viability of human communities by destroying the health and viability of ecosystems on which human health and survival ultimately depend. The whole war system is now a threat to future human survival.

We are learning that human well-being is inseparable from the health of the whole Earth community, and that the healthy functioning of the Earth community is essential to true peace and security. However, few people understood this at the end of World War II, when new structures were being created to ensure greater peace and security that are still largely in place today. At the end of World War I, the consciousness and vision of peace that informed the League of Nations was somewhat limited to the security of states and preventing armed aggression between them. At the end of World War II, the failed League of Nations was replaced by the United Nations, in which the vision of peace was extended to the security of peoples as well as states and included economic development and fundamental human rights and freedoms for men and women. But it was still a largely homocentric vision of peace and security. Nature was not included in the planning for peace and security.

Since then, environmental security has emerged as a major issue, but it has been addressed in bits and pieces. The United Nations Environment Program was created in 1972, following the World Conference on the Human Environment in Stockholm, with a mandate to infuse environmental understanding throughout the UN system, and to undertake educational programs and prepare conferences and international agreements that promote and protect the environment. But ecological thinking is still not at the core of global thinking and planning for peace and security.

The UN Peace University in Costa Rica could enhance that: It could add to the development of that thinking.

PM: Every university should do this. Every university should prepare incoming generations with the tools they need for the twenty-first century. Unfortunately, most universities largely teach for the past, or to get a job that may go out of existence or become irrelevant within ten years. Most do not prepare people in a deepened consciousness of their being part of the Earth and how to help shape vision and policies that will contribute to planetary peace or ecological health.

A lot of progress was made in shaping environmental policy after the Stockholm Conference. But under this Bush administration, much of what was accomplished in the last several decades is being undermined. And after the attacks on September 11, the administration made it clear that the top priorities for U.S. national security are military. It is not just that environmental issues are being ignored. There has been a systematic effort to dismantle environmental policies and legislation that had been made in the seventies and eighties. It is symptomatic of the lack of ecological consciousness to say that environmental protection obstructs business and economic competition, or that there is no solid evidence that environmental problems are real; they are exaggerated claims or hypotheses.

Your awareness that wars start when we lose contact with the Earth is more relevant than ever.

PM: In his book *The Great Work,* Thomas Berry* asserts that the future will be determined by the outcome of the struggle between two competing worldviews: one that sees the world as *commodity* and the other as *community.* Part of the great challenge before us is not only to overcome the war system—our willingness to kill others for oil or water—and replace it with a peace system—systems for equitable distribution of essential resources; it is to overcome a worldview in which we see the world as a commodity that we can sell for a quick profit. We need to embed in its place a deep awareness of our integral relatedness to other beings in one life community. We need to see other beings as subjects in the life system, and not as objects to buy and sell in a global marketplace.

And that's why we can't be happy?

PM: We can't be happy selling ourselves. When we sell the Earth and betray our integral relationships with other beings within the life community, we sell and betray part of our own essence. We are not apart from the Earth, but instead are an inseparable part of the community of life.

You have said that Global Education Associates seeks a balance between the outer life and the inner life for a peaceful and fearless society.

PM: One of the beautiful things Gerald and I were trying to do through GEA was to create a global forum for dialogue between people from

*Thomas Berry, *The Great Work* (New York: Bell Tower, 1999).

different cultures and belief systems on the kind of global structures needed for a more humane and ecologically sustainable future. This process included analyses of world events and trends from different cultural and disciplinary perspectives. But we also explored inner dimensions of global change. What is happening in our spirit and soul informs how we behave in the world. And events in the world affect our soul and spiritual well-being. There is no successful path to global transformation that does not include an inner dimension.

I used to get impatient with people who felt that all the world needed for effective change was for more people to meditate. Many meditations seemed to focus only on personal peace by tuning out events in the world. [Those who meditated in this way] couldn't function in the world because they couldn't sustain inner peace there for very long. Meditation was their way of escaping. They needed to learn how to carry that peace into effective action in the world. On the other end of the spectrum were activists who saw no place for meditation while people were hungry or dying. There was too much urgent work to do to take time for inner peace. Actually, both are needed, and it is important to find some balance between the two. But it is very hard to maintain the balance. We can easily get involved in one or the other. (Laughing) And meditation is so lovely that once you're engaged in it, you don't want to go into the world anymore!

And then the world is so fascinating that you forget all about meditation.

PM: Or so demanding, because you have to raise the money now or your organization will go out of existence. Or, this political issue is being voted on tomorrow, so we have to stay up all night and get everyone to call his or her legislator.

This is a real challenge, because in other times, the monasteries harbored and protected you from the world while you were praying or meditating. And now, being in the world with the same inner attitude is an enormous challenge.

PM: That is absolutely true. When I first came here to teach, I began each class with a few minutes of meditation or quiet time. I needed it for me—to center myself and give my best to the class. But I realized that some of my students did not know how to do it or felt it was out of place. So now I try to center myself before I go to class. I find that many young people are very hungry for spirituality but are not nurtured in their search. Many students here hide their spiritual interests because they feel they will not fit in if others know about it. The modern world is so cynical.

(Sighing, followed by a silence) Religion is so suspect and spirituality is not accepted in a lot of places. It is a constant challenge, as you said.

Why did you leave Global Education Associates?

PM: I didn't really leave. My husband died, and it was very, very painful to work every day in a space we had previously worked in together. He was not only my husband and lover, but also my best friend, my soul mate, and my colleague in very meaningful work. He died unexpectedly and I was totally unprepared for a loss of this magnitude. GEA was in debt at the time. Jerry had done the work of about three people, and was not taking any pay. How could anyone else be asked to do all that he did without pay?

After the funeral, the board decided, in my absence, to make me the new president of the organization. Later, I learned that it had been the board's intent that I provide the intellectual leadership and be freed to write and speak for GEA. They were going to hire someone else to do the day-to-day administration. But nobody told me this. At any rate, there was no money to hire another person, and if the organization was going to go forward, someone needed to be in leadership who understood the mission and work of the organization and could also fund raise. I had been virtual codirector with Jerry since GEA's inception and had been doing much of the writing, editing, speaking, administration, and fund-raising. So, I agreed to be president for a year, after which I would decide whether or not to continue. I continued for three years. But it was tough. I worked eighteen hours a day, seven days a week, trying to do both Jerry's job and mine. I had a lot of physical support from a good staff and wonderful volunteers. But it took all my energy and time to stabilize the organization and move it forward. After three years, GEA was in very good financial and programmatic condition, but I was not. Six weeks after Jerry died my youngest brother also died suddenly and unexpectedly. My family looked to me to help his wife and young children cope, because I "had been through it myself." To others, I must have looked like Wonder Woman, but inside, I was dying. I had taken no time to grieve. Spiritually and emotionally, I was running near empty. I needed time to heal and reconstitute my soul.

Then I was invited by the International Peace Institute at the University of Notre Dame to be one of their visiting research fellows for a year. I took a year's sabbatical from GEA to do this, hoping that it would give me time not only to fulfill a book contract, but also to breathe and heal. But before I got there, two of their faculty went on leave, and I was asked to fill in teaching peace studies courses. Even though this meant less time for my own needs, it was good because it brought me in contact with young adults from places in conflict around the world—Palestine, Yugoslavia, Colombia, Somalia, China, Bangladesh, and more. It was wonderful to work with them and made me want to go back into full-time teaching.

I realized in that year that I believed as much as ever in the mission and vision of GEA, but I didn't want to go back to administration and fund-raising, at least not yet. Meanwhile, I was offered the position of Lloyd Professor of Peace Studies and World Law at Antioch College, and that is how I ended up here in Ohio. It had not been part of my life plan to be in Ohio, but here I am!

(Pausing, then more soberly) I feel history is asking something really, really important of each of us. We should be courageous about responding. We really have to think about what the planet is asking of us and try to find an adequate response. Each one has a job to do. If we commit to do what the planet and history ask of us, I believe the resources to do it will be made available somehow.

I believe that I am being asked to do something new. Recently I started having an unusual dream. I don't dream very often, or at least I don't remember my dreams if I do. So when I wake up remembering a dream, and the dream repeats itself over many nights, it is significant.

The first night after my husband died, something remarkable happened that was replayed in this dream. The day my husband died, I had to face a thousand details in notifying family and friends, arranging a funeral, and more. Then I went into Jerry's and my bedroom and sat on the edge of the bed. I just sat there in the dark, feeling very lost and wondering, "Now what?" Then I turned to look at the side of the bed where Jerry usually slept. From his side, so empty, there began to arise an incredible light that kept expanding and growing brighter and brighter until it lit up the room. I was sitting there, fully dressed and awake, watching this in amazement.

I am a very rational person and do not generally believe tales of the paranormal. I look for a rational explanation for everything, but there was none I could think of for this experience except Jerry's powerful presence. He was telling me that he was in a good place and that I was going to be all right.

Then, a few months ago, this incredible light showed up again, but this time in my dreams. At first, because I associated this light with Jerry's passing, I thought my own passing was imminent. Interestingly, I was not frightened, but instead felt very calm and reassured. I had been going through some very painful experiences, still working through the loss of Jerry, but also with some turbulence in GEA under new leadership. I had to either let go or resume leadership. I had written a number of letters of resignation but never sent them. I began to think this brilliant white light appearing in my dreams might be a symbol not of impending death, but of returning life and creativity. The long loneliness and pain of losing Jerry and my sense of self were giving way to a sense of inner calm. With each dream came the mantra *Follow the light.* About this time, at least three visitors who had not seen me in some years said, out of the blue, "What are you doing here? You belong in a bigger arena, doing something more meaningful and important." But nobody could tell me what that was or how to find it. I feel that there is something new or different that I am supposed to do and that it will be clearer to me in due time.

Probably you will always be teaching, in some sense.

PM: In some sense, yes. Isn't that what we're all about? We are all learners and teachers.

It is very obvious that you are among the pioneers awakening people to a new way of thinking: that we are part of the Earth and that people need to understand that. You are showing people the way.

PM: Yes, but I don't feel as though I have to be in one place to do it. What I mean is, it may not matter whether I do that here in Yellow Springs, Ohio, or somewhere else. I have heart homes in a number of places around the world—Africa, the Philippines, Japan. I am not place-bound, but I do feel bound to content and to a mission. And I have to think about what will take me to the next step in that work . . . .

That sounds exciting!

She responds with a laugh that stands midway between excitement and restraint, as if she suddenly realizes that by saying these words out loud, she is making a promise to herself.

Let's talk more about education. Are you familiar with Jane Goodall's work with kids all over the world in her Roots & Shoots program?

PM: Yes, I'm familiar with her work and I was really taken by her last book, in which she shares her own spiritual awakenings and growth. I have been a speaker on programs with her and had an opportunity to sit next to her and talk once, as we're talking now. She's an amazing woman, very humble and courageous, with great inner strength.

You have talked about cosmogenesis. Can you explain what that is?

PM: Thomas Berry introduced that term to me. The cosmos is not in a finished state. It is still evolving, a work in process. And we are part of the process. Our choices influence the further stages of evolution, at least on this planet.

We could take that even further: In our thoughts and in our decisions, in what we do and what we don't do, we influence this evolution. It is not only the big decisions that are important, like the ones that are made during the Earth Summit; it is also important how you treat your children or your neighbors and yourself.

PM: We create our own souls by the little choices we make, and we can decide to honor the larger soul by making choices that honor the community of life. All people together create something like a collective center contributing to the soul of the universe.

I reflect on how one of the problems of our time is that instead of knowing that they are part of the creation of our common future, people underestimate themselves; they think that they don't matter. It starts with an educational system that wants kids to unlearn what they already know and then stuffs them with a lot of ideas and facts they don't need. We could educate the young to be aware of what they do know and stimulate who they are. Above all, we can help them to understand that they know a great deal, that they matter, that they can make a difference, and that it helps to learn to really listen to others. We teach each other. Much of this has to do with what Patricia has already said: If we are disconnected from the Earth, we are disconnected from ourselves, and vice versa. This disconnection causes us to feel that we are not seen anymore, do not count anymore. And there the conflict starts. The most common cry of people on Earth today is that they feel overlooked.

Patricia tells us about the recent birth of her grandchild, Rachel, and how incredibly happy this baby was about her arrival. Interestingly, the first sounds from little Rachel were not a cry. She came out babbling and cooing, full of energy, almost talking her way into life. And then Patricia's eyes, which had sparkled when she was talking about her grandchild, turn suddenly dark as she continues.

PM: Babies are like seeds that carry all the potentiality of the Earth in them. Those seeds may grow or not grow; the conditions may be right or not right for that seed to become a giant tree. The great tragedy in so many lives is, as you say, that they never receive the love and nurturing they need to grow. Each new child is born with all this potential, and then so many horrible things can happen to them, so much damage can be done. I am still stunned at how much violence has been done to young people.

Older generations have responsibility to help the next generation unfold in a way that [helps] them and the Earth reach their full potential.

Do you think we are the last species?

PM: Oh no! I mean, I don't know. But I hope we will survive and not become one of the extinct.

But then something else would come after us.

PM: Well, I hope they look more like us than like cockroaches!

Some people say cockroaches are beautiful . . .

PM: I can't find it in me to call them beautiful. When I lived in New York, it was a constant battle to keep them from cohabiting our apartment. I never learned to love them. But I do admire them for their tenacity. As a species, they have survived longer than most other beings.

This talk of species survival brings to my mind the Mayan calendar set to end in the year 2012. Some scientists say that we have only ten years left—that between now and ten years from now, the sixth extinction will occur, much like the disappearance of the dinosaurs. They say that only some insects and viruses will survive.

PM: If you think of evolution in a different sense, the good news is that while individual lives are lost, life as a whole has always found a way through impasses on the evolutionary road. As Thomas Berry notes, life is full of surprises and creative solutions. Early in cosmic evolution, after the big bang, when forces of antimatter had destroyed most of the matter, a few particles found a way through, and out of this came the entire universe. And early in the evolution of life on Earth, when cellular life had consumed all the nutrients needed for life to continue, photosynthesis was invented, and from this came the greening and flowering of the planet and an incredible diversity of life. The human capacity for creativity is also impressive. Obstacles and crises do not need to mean an end to human life. They are also opportunities for us to generate new directions and grow in greater life and maturity.

It is running late and jet lag is slowly creeping up on Jessica and me, so we decide to end our conversation here, knowing that tomorrow morning we will have time to continue to talk about so much more.

The next day starts with the American custom of sharing breakfast with a group of people. Patricia has invited students and professors who are interested in talking to me about my experience in the field and, of course, about this book. After this visit, Patricia, Jessica, and I sit down one more time.

In the introduction to your latest book,* we can read about the theory of different time frames of analysis for the future: fifty-, five-hundred-, and five-thousand-year time frames. Where are we now?

PM: At a time of change, however you look at it.

Can you explain your theory?

PM: What I say in my introduction to the book is that we are in the middle of a time of tremendous change. The question is not whether the world is changing, but how deep this change goes. There are a number of different levels of analysis.

One analysis covers a fifty-year framework from the end of World War II and the founding of the United Nations. The United Nations was founded to prevent another world war and to resolve problems of peace and security as they were then understood and within the framework of sovereign nation-states. Since then, the world has become more economically, ecologically, sociologically, and politically interdependent. National sovereignty has been seriously eroded in a rapidly globalizing world of new regional arrangements. The United Nations isn't an adequate framework for these new global forces. Many transnational corporations now have more power and wealth than many nation-states, and they exist outside the UN framework. The United Nations is a state-centric organization, whereas many powerful actors today are not nation-states. Nor was the United Nations given sufficient means or

---

*Patricia Mische and Melissa Merkling, *Toward a Global Civilization? The Contribution of Religions* (New York: Peter Lang Publishing), 2001.

authority to protect the global common good. It can be held hostage by the veto of a few powerful nation-states. In the twenty-first century, we will need to greatly reform and strengthen the United Nations or develop new, appropriate systems of global governance to manage global-scale problems and ensure peace and security and the global common good. So that is one framework of analysis.

Another analysis takes a five-hundred-year view of transformation and considers the inadequacy of the nation-state system in the face of challenges from above and below. The nation-state will continue to exist, but its power will continue to be challenged by new global actors that cannot be contained within a nation-state frame. These actors include everything from transnational corporations and religious networks to intergovernmental regional organizations. They also include terrorist networks. Among the most hopeful signs is the emergence of a global civil society—citizen groups working in solidarity across national lines that, in the absence of effective action by national governments, seek to shape global policies from below.

But there is another, third level of analysis. It agrees with the other two but goes even deeper. It examines a five-thousand-year framework dating from the rise of the war system, class system, racism, sexism—all part of a paradigm of dominance over the Other—and a worldview in which humans are dominant over the Earth. All of these systems of dominance are now being challenged, but the most difficult to get at is the system of human dominance over the Earth. The myth of this dominance is deeply embedded in the Western psyche. We see ourselves as separate from and above the Earth, rather than as part of a community of life. But more and more people are challenging this myth and supplanting it with a vision of humans as part of a community of life. This is a very, very big transformation that ultimately requires transformed consciousness as well as politics.

What do you see happening in the world? You've said that we haven't yet understood that we're part of a community of life on Earth. That means we are at the beginning?

PM: As a human species we are [unaware], although some people are very far ahead. People like Arne Naess, with his deep ecology philosophy, and Tom Berry, with his new cosmology, and many ecofeminists are well ahead, but in general we are at the beginning.

The work of Aldo Leopold on a "land ethic" in the 1940s was revolutionary at that time. By *land,* he didn't mean just the soil; he meant the total biological community, the community of life. He said that we had to extend ethics to include the whole community of life and with it a proper understanding of the interactive dynamics of the members of the community.

Is feminist ecology based on the idea that women are more linked to the Earth?

PM: People used to think of the Earth in feminine terms.

What is the role of the individual in all this?

(Placing her hands on her forehead and her two index fingers on either side of her nose in a gesture of concentration) I think that the question of individual human choices and human consciousness is really critical. We can think of this in a number of ways. From the viewpoint of the Earth, the emergence of human beings and human consciousness and choice was an event of extraordinary proportions. As Thomas Berry has noted, humans were from and of the Earth, but they were the Earth in a new way. Through humans, the Earth was able to reflect back on herself. This is an extraordinary way to view the importance of human consciousness in the collectivity.

But the question of individual human agency is also very important. We are beings with free will and choice, and capacities to think at very complex and sophisticated levels. These choices can sometimes have a significant impact, not only on our own lives, but also on the further stages of planetary evolution. This means we have extraordinary responsibility to use our powers wisely on behalf of the good of the total planetary community. We are co-responsible with the Earth for the further stages of evolution. We haven't yet learned to take responsibility for our choices. There has been an ethical lag in our development. We need to grow up and become more morally mature.

Then we have to grow up quickly.

PM: I think that it comes down to each one taking responsibility for her or his own life. Because each life does make a difference.

You have written somewhere, "We have to redefine what it means to be human." That is so beautifully said, and so true. I would like to hope that we are moving on to a broader awareness.You have written much about religion. Does it in any way fit into all of this? And is the role of religion getting more or less important in this changing world?

PM: Religion in the sense of "seeking meaning" is still very, very important. The spiritual core of the great religions is still very important. But the structures around it, the institutions, have to go through continual renewal to stay true to their spiritual core. Religions are enormously significant in that they often carry the archetypes and symbols of meaning for a culture. They also have practical importance in bringing together people in community around a spiritual vision and norms of behavior. Religious networks can be very powerful global actors. I have seen this in work for human rights and peace, and more recently environmental action. Buddhists, Hindus, Christians, Jews, and Muslims and other religious networks are now engaged in many forms of dialogue and collaborative action. Working together, they can be a powerful force for positive change. I have an enormous respect for interreligious collaboration. I grew up in an interreligious family. But, perhaps because religion is such a powerful force, it is sometimes used by ruthless or unthinking people to justify or do harm. This has to be guarded against through constant effort and renewal.

I see religious networks as one of the few forces available to challenge abuses of state power. Thus, it is unfortunate when religions become synonymous, indistinguishable from, or used by the state.

It is interesting that Christianity, of all the religions, has been separating itself most from nature. Or don't you agree with this?

PM: (Taking time to formulate her answer) I don't know whether I can agree. There have been many Christian thinkers and leaders who have done quite the opposite—taught about the deep connections between ecology and Christianity—people like Meister Eckhart, Francis of Assisi, Hildegard von Bingen, and others. There are different streams in Christian teaching, and at different times, different schools of thought have been more prominent. There is a major resurgence among Christian groups now about relationships between ecology and Christianity.

I don't know whether you have seen the wonderful work of Mary Evelyn Tucker. She and her husband have done a lot of work as professors of comparative religion. She also worked with Thomas Berry. They explored what is at the essence of religions. They came up with . . . the relationship with nature! That's when she said, "Let's bring them all together." Thomas Berry, who was also a scholar in comparative religion, believes that all religions initially emerged out of a deep sense of relationship with nature. The spiritual impulse may have arisen from a primal sense of awe and reverence for nature. The original religions were all nature based. It is important that members of religions examine the roots of their respective traditions.

Thomas Berry has observed that each religion has important contributions to make toward the development of new relationships with nature, but none has a complete picture; none is adequate in itself for the kind of transformation in consciousness we now need. We cannot just go back to old, traditional understandings. But we can bring the best insights from each tradition forward for what they can offer as we try to develop new patterns of relationship with the natural world. However, we also need insights from science and a proper understanding of the cosmos and planetary dynamics.

The historian Arnold Toynbee studied all the great civilizations and came to this conclusion: Religion and spirituality played a far more important role in the rise and fall of civilizations than most historians have acknowledged. New civilizations arose around a spiritual vision held by a creative minority. The real shapers of culture and civilization were not political leaders, but spiritual leaders such as Buddha, Abraham, Moses, Jesus, Paul, and Muhammad. Civilizations that lost their spiritual core fell into decline and died. Then, from the ashes of the old civilization, there would arise a new creative minority who would organize around a spiritual vision that would serve as a chrysalis for a new civilization.

After World War II, when the evils of fascism were raw in people's minds and atomic bombs began to pose a new threat, Toynbee warned that nationalism was a kind of false god and a dangerous substitute for authentic religion. The state was assuming godlike powers over life and death. Authentic religion and a renewed spirituality were needed to tame the threats of excessive nationalism.

Is that what is happening in the United States at the moment?

PM: Absolutely. This has not happened to the United States on this scale since World War II. You see, feel, and hear "God bless America" everywhere. You would almost think God is America and America is God. Everything we decide is right and if people don't do it our way, they must be evil!*

I feel lately, seeing U.S. flags everywhere, even printed on shirts, shoes, towels, baby's diapers, and everything else, that the flag is a symbol of a belief system, a kind of religion, and that Toynbee was right to warn us against the dangers of nationalism as a false god.

But the cynical aspect of this is that this whole glorification of the flag may stem from our fear of the religion of the enemy.

PM: Yes. It is as though it's our religion against their religion, except that we don't say it's our religion—but it is our belief structure.

I think that all of this will, in the end, lead to spiritual search. Most people are not satisfying this spiritual hunger in church, although maybe some do. At a fundamental level, some people may need a religious authority to tell them right from wrong, and [they believe] that if they follow this path, everything will be okay. People don't want to take

*Interestingly, Patricia shared this view months before the United States and Great Britain started the second war in Iraq and before George W. Bush openly stated that all countries that weren't willing to fight with the United States were considered to be against the United States.

personal responsibility. But underneath this, there is a great spiritual hunger; I see that with my students, but they can't find a community in which to develop and nurture this part of their essence.

I see this spiritual hunger and thirst in people all over the world. If Toynbee was right that spirituality is a driving force in history, then is today's spiritual quest an indication that we are at the beginning of the rise of a new civilization—this time, a global civilization? We need to delve deeply into this spiritual search, because nothing we have to date is adequate for the global scale of the challenges ahead. Nothing we've known up to now is adequate. We are on new ground, in the process of a major transformation. Hopefully, there are ways we can reach out to each other and strengthen and assist one another in this search. We humans are the new creative minority all are waiting for.

We need to find a force that is more powerful and authentic than the false promise and belief in "my country, right or wrong."

And more fulfilling!

PM: (Laughing loudly in agreement) I think we are in a time of global transformation, and that spirituality now has to be global. We need a spirituality that stands in reverence for the whole cosmos, not narrow nationalisms. If it is not that big, then it is not going to be big enough.

But how do we meet these challenges with political action as well as spiritual growth? We have to integrate new cosmology and consciousness with effective political strategy, something comparable to what Martin Luther King Jr. did in the civil rights struggle. To be effective change agents, we have to affect policy and therefore we need a political strategy. Some would say that education for a new consciousness is strategy, and yes, it is. But we also have to concentrate on effective political change, or all our spiritual consciousness may be ineffectual.

You say that in working so hard, you are losing the sense of the oneness of life. But is that true? Don't forget that you are intrinsically bringing nature into your teaching, which means that you are connected.

PM: I walk five blocks to school through an area that is not highly urbanized. I feel that I walk through the changes of the seasons, even if it is only those five blocks. And this village is extremely beautiful in spring . . . Maybe you're right that I do keep experiencing it; even if I'm not devoting time to it, it comes to me! It is still there, and then I know that things are working together, and that I am part of it.

Do you feel that you are a pioneer in your work and thinking? And if so, do you feel the loneliness of a pioneer?

PM: Sometimes it is lonely. But then I find other people to share it with. I feel that everywhere around the world there are incredible people, and when I'm with them, I don't feel lonely anymore. But if I don't give time for these nurturing experiences, I can feel lonely. I need regular communication with others and with nature.

I used to travel a lot for GEA, on speaking programs, and those trips were very nurturing because I was constantly learning from the people I was meeting. But there is a danger when we work in the same structures, such as universities and organizations, that we stop being challenged.

Patricia, what is love?

PM: Love is the strongest power in the universe. It is attraction, bonding . . . It is what holds things together and makes life possible.

Just then the doorbell rings and a friend Patricia has invited arrives for a walk. She wants to take us to an old forest nearby. We will finish our last question surrounded by the mysteries of the woods, under Yellow Springs' lovely old trees, where all is silent. We reach a spring with water rushing clear and beautiful. We think we suddenly understand the meaning of the name of this town—yet the stones and rocks beneath the water are not yellow but unmistakably orange! Patricia's friend explains that they made a mistake when translating the name from the Indian dialect. In the Indian language it was called Orange Springs.

Here in this place filled with abundance, we feel thankful for all this life around us and for our exchanges these past two days. I ask Patricia again.

What is love?

PM: I said earlier that it is the strongest power in the universe in the sense of attraction and bonding. But I think it is also recognition of the sacred in the other. It is a response from one's inner being to the essence of the other being. It is like leaping over reason, intuitively uniting with that being, with the sacred and the essence of the other being. And it is respecting the other being as distinct.

Works by Patricia Mische

Ecological Security and the United Nations System. location: Global Education Associates, 1998.

Star Wars and the State of Our Souls. Minneapolis: Winston, 1985.

Toward a Global Civilization? The Contribution of Religions (Patricia Mische and Melissa Merkling). New York: Peter Lang Publishing, 2001.

Toward a Human World Order (Patricia Mische and Gerald Mische). New York: Paulist Press, 1977.

EPILOGUE

My desires in conversing with the twelve individuals in this volume were first to learn from them myself and then to offer their thinking and experience to you, the reader of this book. It is my hope that this has allowed those who are familiar with only Sheldrake's work, for example, or the words of Jane Goodall, to meet the others I've introduced. Perhaps this will encourage you to explore the thoughts and ideas of these twelve beyond these pages, and to explore the fresh ideas and inspiration brought to you by both the connections and the differences among them.

I also hope that you have learned more about the variety of endeavors that people pursue with and for the Earth. Above all, I hope the stories told here will bring you fresh ideas and inspire you to rethink and reevaluate your own relationship with the Earth and your place in the web of life. How deeply will you change your way of thinking? How seriously do you take yourself and this endeavor? Do you love the Earth and yourself enough to embrace not just the awareness but also the responsibility of being part of the beautiful symphony of life? Are any of us truly up to the task? The truth is we are all responsible for our common future, the future of the planetary community.

There is so much to do! Enlightened public policy is necessary, as are national and world conferences, but ultimately it is our actions, both as individuals and collectively, that matter most. We are, after all, members of the great community of life. Each of us matters in the smallest choice we make, in the way we think and act.

The actions each of us takes influence not only the people around us, but all that live in our proximity and beyond as well—just as our surroundings in turn influence us and our actions: We know, for instance, that crime is more prevalent on streets without trees than on those lush with greenery, and that there is less criminal activity around shopping malls surrounded by trees than around those surrounded by nothing more than concrete. The material as well as the nonmaterial value of our environment matters.

The greatest imperative is that we respect and sustain nature in order to preserve life on Earth for our own children and grandchildren as well as for the children and grandchildren of the world's elephants, birds, trees, and butterflies. As part of the whole, we can no longer disassociate ourselves from in integral *sustainable development,* a concept that is growing in the mind of the public but not necessarily in the mind of the policy makers.

There are various definitions of *sustainability.* The one I like most defines it as "the growth and development of each individual without curtailing freedom of choice for other people or groups." I would add to this: "without curtailing freedom of choice for other people or groups or *any other living species.*"

It is clear that our treatment of the Earth today is not sustainable. Credo Mutwa considers the outlook bleak, given the harsh reality of the Earth's condition. Is 2012, the Mayan year of decisive change, lurking? Are we indeed on the verge of the sixth extinction? And how will that present itself? How do we feel about these visions of the future? How do we position ourselves and humanity as a whole in the future of the planet? Are we conscious that perceiving the future as a threatening specter only reinforces its negative power?

Perhaps the major change in 2012 will involve a significant cleanup in our minds. Having formed our current society through our thoughts and actions, it is our responsibility to repair what we have broken. If we aim our intentions on restoring this beautiful planet, we should be able to reverse our actions for the well-being of the Earth. Each of us has an important role to play here. Perhaps we cannot live without creating some form of pollution in our environment, but each of us can certainly reduce the amount of waste we contribute to the air, earth, and water. There are many subtle ways to make cleaner both our personal habits and our habits as a group.

This brings us to Rupert Sheldrake: Treating ourselves and our fellow humans with respect can only have a positive impact on the Earth. Creating a positive morphic field will benefit everyone as we work to live in a way that is considerate of the Earth, others, and ourselves. We can all do this every day—though remember Sheldrake's words: "There will always be Eeyores to strike the balance." We must also remember this truth: All life-forms live in the now; we humans are the only ones who don't, for our thinking gets in the way.

With regard to sustainability, Jane Goodall's vision is refreshing because she replaces the concept of sustainable development with that of a *sustainable lifestyle*—which requires our inner development, an awareness of how we handle ourselves and our surroundings. Today, the term *development* usually refers to economic growth. But how much can we grow economically without further taxing the Earth? Economic development—which should be, first and foremost, the struggle against poverty—has to involve personal development as well as ecological development and preservation of biodiversity. According to Arne Naess's philosophy of "deep ecology," we as a people are part of the life process, which gives us both benefits and burdens.

Yet we *can* take the next step toward healing our separation from the web of life, which goes hand in hand with healing the Earth. Gareth Patterson says:

> Destruction of the natural world can be recognized as self-destruction. Acknowledging this as a fact, we can start to heal ourselves and the natural world. The health of our planet and our own inner health are one. The more we took from Earth, the more impoverished we became; nothing is meant to be separate from the whole.

The Mayan culture, represented in this book by Rigoberta Menchú Tum, lives with this integral view of life. She says: "The heart of Earth is like our mother . . . who nurses her children, who nurses life." Similarly, Matthijs Schouten says: "Sometimes I think that separation is the only real crime, from which all suffering originates." He points out that our ecological problems force us to think about who we really are and what we are doing here. To him, it seems logical that our daily practice cannot be separated from the spiritual, the oneness with all; nor can we separate our personal, inner work from practical, even technological, solutions.

As I see it, reevaluating our relationship to and role in nature requires both a practical and a spiritual approach. Spirituality involves respect for the essence of every species. Spirituality embodies connection and interconnection—to our inner selves, to all that is here on Earth in its materialized form, and all that exists in nonmaterialized form.

When I pose to Rigoberta Menchú the question "Are we the most important species in the world?" she looks perplexed. Her subsequent response is very revealing: "Ask the Earth." The wholeness of all life is so much a part of her way of living and thinking that she cannot understand the idea of separation or fragmentation, of most or least important: "Such fragmentation is bad for life, as it enters impossible and inconceivable divisions in everything that should be regarded as indivisible." She urges us to take the time to look at what we have done to the world, and to take a good look at our inner selves: "We need to invest in nature and to restyle our behavior."

Credo Mutwa also speaks to union and spirituality when he warns us: "Expelling God from everyday life leaves the field clear for the super capitalist, the colonist, and other plunderers to rape the Earth, to destroy nature, to rape priceless natural resources with cold impunity." He asks for the human spirit to rise.

Today, we are witnessing a great ignorance regarding all the Earth's life-forms, ourselves included. Education seems necessary to raise awareness, to open our consciousness to feeling and an awareness of our own deepest essence as part of nature, at work as well as at home. All life has intrinsic value. We are here to experience our own value and from this to give to all else—whatever this may involve. We are here for the whole of life, for we are part of that whole. Arne Naess brings us this reality through the eight points of deep ecology, and Denise Linn says: "I believe it is damaging for the soul if we separate ourselves from nature. Separation between man and nature and separation from our own inner nature go together."

Rupert Sheldrake expands upon this idea of division from our inner nature. He points out that it is not so much that we should love nature, but rather that we should first heal our separation from ourselves, the division between the person we are at work and who we are at home. "What people think [of as] their 'true self' is not the person they are in the office. People think their true self comes out on holidays, weekends, and after working hours. But [the two are] the same person . . ." Healing the separation within ourselves is imperative for healing the Earth. Goodall characterizes this separation as the scientist who puts on his white coat and tortures dogs in the lab, then at home, without the white coat, says " 'My dog understands everything I say.' That shows there is a split in the brain."

Patricia Mische also addresses the connection between our own health and the health of the Earth: "We are learning that human well-being is inseparable from the health of the whole community and . . . the healthy functioning of the Earth community is essential to true peace and security." She adds: "There is no successful path to global transformation that does not include our inner dimension. There has been an ethical lag in our development; we need to grow up and become morally mature."

James Wolfensohn, who has wielded the might of the World Bank, focuses on justice and on access to health care, education, and property rights for women and children around the world. In addition, his work has been dedicated to a sustainable future that can relieve the pain of those living in so-called developing countries. Through education, this effort toward making a sustainable future a reality also surfaces in the work of Jane Goodall, Patricia Mische, Matthijs Schouten, and, in some respects, Rigoberta Menchú. In her Roots & Shoots program, Goodall reaches out to children by gently alerting them to their responsibility to both society and their surroundings. She dreads the moment when these children grow up and see the world as it is. But they will be stronger for having learned through Roots & Shoots, and they will stand tall in this world in a way that is just.

Mische's peace education guides students who are still searching for their place and responsibility. "The whole war system is now a threat to . . . future human survival. Ecological thinking is still not at the core of global thinking and planning for peace and security. We need to prepare people in a deepened consciousness of their being part of the Earth and [a sense of] how to help shape vision and politics [that] will contribute to planetary peace and ecological health."

In his lectures on ecology and restoring nature, Schouten educates by combining philosophy with biology, uniquely inserting the diction of healing into the academic dialogue. He can be seen as a link between the scientific and the ancient and Buddhist approaches to nature. Rigoberta Menchú teaches as well, including ancient knowledge in her book of fairy tales so that children all over the world can become aware of the wisdom of the Maya and gain a solid foundation to build upon in the future.

To all of my discussion partners I posed the question "What is nature?" To me, Rigoberta Menchú's perspective, her femininity, and her way of living and being represent the answer: Nature is all of life, and no form of life can exist without the other species that live on our Earth. Menchú lives this unity—and the diversity within it tells of the magnitude and power of the Primal Source. Although we may be unable to fathom this Source, we can feel it and we can live connected to it, for as Rupert Sheldrake says, "Nature is the entire universe."

I consider nature in a very broad context that includes our inner nature or essence. Our approach to our work, our interactions with each other, and an honest, clear way of dealing with ourselves should result from this starting point. In this view, it is easy to see how helping someone cross the street or paying attention to what we eat to nourish ourselves can affect not only ourselves as individuals, but all life around us as well.

If we place ourselves within nature, within the flow of all life, we become aware of the overall interconnectedness of everything that is alive. This, in turn, allows us to live a natural, connected life, which is entirely different from one that is lived in seclusion or even unconscious separation. This connection gives each of us the power to effect change. Masaru Emoto has said that if 10 percent of all people become aware of their choices and opt for a healthy world, our society will change. Here again, according to this view, we see how the actions of each of us matter and that each of us is a unique individual connected to the collective.

Rupert Sheldrake relativizes this position, however, by saying that if we move a cup, the change will indeed affect everything around it, but not vastly. This important distinction adds some perspective: We need to take our responsibilities seriously, but we cannot let them suffocate or paralyze us. Those who are able to keep up their spirits despite the world's problems will make conscious living easier for others. This understanding is, in itself, a major step toward coexistence, which is what nature both expects of us and shows us.

Viewing ourselves as part of nature also involves rethinking an important question: Do all living things have awareness or *soul?* It is essential for us to understand that everything *lives*—this flower in the field, that tree in the woods, this bird in her nest, or that ant crawling across your leg is alive and, as a consequence, feels to some extent. But does it have feelings or emotions?

Of course, we know that a flower has life—but are we truly aware that it is *alive?* Accepting that all life is truly alive, that everything living has a certain awareness, becomes easier once we understand that a plant, for instance, feels our presence. Knowing this, we will treat the plant differently. Arne Naess says that what we acknowledge as living will come alive for us personally.

Is having awareness the same as having a soul? The professionals interviewed for this book have different opinions about soul and awareness in all living things. Some believe only living organisms that think can have a soul. Some believe thoughts can exist only if awareness exists. Rupert Sheldrake explains that he prefers the term *soul* to *awareness* with regard to the animal and plant kingdoms—but he attributes potential awareness (and thus thought!) to the sun.

Regardless of individual opinions, the discussion of soul and awareness in all of nature is important because it encourages us to regard all life as deserving of our respect and consideration.

&

We can say that overall much has changed in the attitude of humans toward nature. Fortunately, people who regard themselves as both separate and "above" the rest of life on the Earth are becoming fewer. Of course, we do not own the Earth. Rather than defining our role as one of ownership, then, it seems far more balanced to term it *trusteeship*. At the University of Nijmegen, Professor Wouter de Groot has investigated how people at this moment in time view themselves with respect to nature and learned that most people appear to see themselves as *trustees*. Some, he determined, consider themselves to be *partners* of the Earth; a select few feel that we are *participants* in the community of life on the Earth; and a seemingly small number of individuals have perceived the unity of everything alive in a "magic moment"—a *union mystica*.*

The study reveals that people have indeed changed their way of thinking and, perhaps most important, the way they feel about their relationship with nature. In my view, our role is that of trustee, partner, and participant combined. As trustees, we understand that we need to water the plants, that animals need enough space to live, and so forth; we can provide for the needs of other living things on the planet. Being a partner goes one step further, symbolizing equivalence. It requires that we listen and open ourselves to living things on Earth as comembers of a group. As participants in Earth's life community, however, we are required to assume a more modest place in the whole. Our role varies among these three depending on the situation. Yet because we are perpetually at risk of separating ourselves from nature and resuming a dominant role as trustees, it may be wisest to regard these three roles as coincident at all times.

Matthijs Schouten best explains the beauty of our role as participants in the web of life:

> A plant can reveal its entire essence to us only when we have discarded both what we have learned about it and any other image regarding it. We will need to accept what is shown to us, even if we do not like it or if it is incompatible with our frame of reference. We

---

*Wouter de Groot, "Natuurbeelden, Eco-spiritualiteit en het postmoderne Zelf" [Images of Nature, Eco-spirituality and Postmodern Self], lecture delivered on October 1, 2002, Studiecentrum.

> must be willing to let what is revealed to us change the way we view ourselves . . . Whoever wants to give space to the fullness of life has to conquer his or her greed, anger, egoism. Imposing ourselves in any way on the subject, we cannot learn from it, as it cannot reveal itself.

As for a magic moment—a union mystica—I found that it occurred for me exactly as Schouten describes: at the very time when I was so tired that I let go of all self-imposed constrictions that the ego carries with it. Only then could I become fully receptive to the "other" as it truly is. Such an experience opens the true face of nature and our nature within. As Schouten says of walking through the bogs in Ireland: "I just was, so we could 'be' together."

Sadly, we don't talk about these beautiful moments of connectedness in and with nature; they are often deeply personal experiences that we don't want to share. But because nature needs our care and attention more than ever before, the time has come to share these moments, so that it becomes normal to speak about both our inner nature and our relationship to nature. The dialogues and stories in this book are evidence that none of us is alone in our heartfelt connections to the Earth and in our quest to restore unity from our separation.

In communities of ancient peoples, coexistence with nature is nurtured—at least on a modest scale. The Kogi Indians in Colombia, for example, have created a society in which personal development is part of the culture. They make room to discuss possible emotional problems and thereby avert or resolve outbursts. Further, the Kogi believe they can be themselves by connecting with all life-forms. In these ways, their culture fosters the connection between themselves and nature.

Denise Linn has felt this connection: "Nature teaches me to be honest with myself and others, to be honest with my feelings and not suppress or deny anything. Nature gently removes me from the confining and defining world, immersing me in stability and interconnection." But it is clear that it is a challenge in today's Western society to find ways to foster this connection on a broader scale. Yet people *are* searching for these means—for a more meaningful life. Materialism has not fulfilled the quest for happiness. Many so-called developing countries are looking for ways to live comfortably without polluting the Earth. In

China, for example, the humanistic neo-Confucianism that people seem to be returning to tells us that heaven, Earth, and all life are one.

Let us start to educate children in this holistic worldview, imparting the sense that they are world citizens—which of course they are—but also, maybe foremost, that there is a community of all life of which they are part.

In my conversation with Rupert Sheldrake, we discussed how puberty here in the West often coincides with rebelliousness and even aggression. He attributes much of this response to a lack of involvement in society at the moment that young people feel they are ready for participation. Sheldrake regards the problem as distinctly Western. How can we resolve it? How can we offer children a chance to be themselves from early childhood?

Many children and adolescents today show signs of being either hyperactive or withdrawn. Some children are extremely sensitive and react to their surroundings in sometimes violent ways. Too many children are being treated with drugs meant for grown-ups. We can ask, "What is wrong with these children?" Or we can learn what is wrong with our society—a society in which so many young people apparently don't fit. We can listen to what our children have to say, instead of forcing them to operate within our current way of thinking. They can guide us, provided we take notice of them. They can teach us to transcend systems, the set patterns that chain us in education, government, health care, and corporate industry.

Most of our reluctance to allow room for those who are different or who think differently arises from fear of change or fear of losing our so-called truth. If we continue to program children with what we already know, however—even when we don't like what we are—we will miss the opportunity to benefit from the new insights and changes they can offer us. Children who are not allowed to be themselves may at some point lash out as a response to being constrained. In such instances, they are quickly classified as "difficult," though it is their sensitivity that intuitively leads them to resist a system they feel is not right. Rigoberta Menchú says: "If the main purpose of our laws is to punish violations and offences, and little is done to prevent them from happening and to avert a loss of respect for others, then the writing is on the wall. It is time to look at what we have done with the world and to take a look at ourselves."

Personal growth, allowing ourselves to be who we are in the core of our nature, is a necessary contribution to changing our attitude toward the web of life.

Many of us wonder if science and technology can save us from the pit into which we are falling or, as I ask in the introduction, if there is a balance between what we call *science* and our personal experience—one that allows us to live in a more harmonious and sustainable way with all of life. Credo Mutwa believes there is: "We can give technology a human face." Matthijs Schouten also believes this balance can be achieved: "The rational way of approaching things is just one way to look at reality—a very important one nevertheless, since it has given us so much. But science should not exclude everything beyond the rational concepts. In the Buddhist view, ultimate reality, ultimate truth, lies beyond rational thought, beyond any concept."

Sheldrake carries out research on disputed scientific topics, pursuing solid evidence. There are vigilante groups, he tells us, that want to maintain the rationalist worldview. Yet it is science that brings him closer to the miracle of nature and of life. Jane Goodall, too, makes room for science: "I don't think science needs to be cold, hard reductionism. Science in its true form is wanting to learn and know about. If you approach nature with a curious mind and a sense of wonder, I think your sense of wonder becomes deeper and deeper the more we learn about it."

Hans Andeweg notices that sometimes scientists have a fear of standing out in ways that could amount to professional suicide, "[w]hereas science, by definition, should continuously explore frontiers and be willing to extend them. Transcending the borders and stepping out of the mold takes courage." Sheldrake gets at the crux of the problem: "The separation of science, spirituality, and the sacred underlies our present crisis of ecological devastation, despair, and disempowerment. How else can we hope a new sense of meaning is awakened, if not by the coming together of those powerful traditions that were sent asunder in the seventeenth century? We need a cosmology that speaks to our hearts as well as to our minds."

Finally, we must consider the answer to the core question posed to the interviewees here: What is love? The answers are profound. Rigoberta offers: "Love . . . means being coherent. There must be coherence between what we say, what we live, and what we do. 'Living with love' means being coherent." The wise Credo Mutwa says: "If we could say the word *love,* the whole universe would dance." Masaru Emoto suggests that "[l]ove reduced to one love and one gratitude is why this world has become materialistic. But the essence of love, of nature, is one love and two gratitudes." And when Patricia Mische speaks of "Earth Literacy," she speaks, of course, of being open to the interconnection of all life around and within us, and is not that kind of openness love?

To me, love is first clarity. It is a vibration perpetually in motion and is therefore intangible. I regard love as the life or light that shines within all species, that forms the connections among all life, which is bound to the Primal Source.

In relationships with our children, with our partners—in all our relationships—love means respecting others in every conceivable way. In fact, it cannot exist without respect. *Conditional love* is what results from love without respect, and it leads to power struggles that arise from greed and desire or from fear of not having or of losing. Conditional love leads irrevocably to suffering by creating division among us. Our feeling of separation evokes fear—but the separation is an illusion; we are not in fact divided or alone. Ultimately, all is one. As we have seen in this book, all who have chosen to be lovers of the Earth have transformed fear into love for life itself.

Humankind might not be able to heal the destruction it has caused without the support and vital forces of every species of the great community of life, just as the illness of one organ cannot be healed without the support of the entire body. The animals and the power of the Earth herself, even of the entire universe, are now involved in this great healing.

Every one of us is thus challenged to open up to the healing within ourselves and to participate consciously in this greater healing, together with vital, vibrating nature around us. Let us open our senses and that deep inner knowledge we call intuition and in the process give each other the respect that is so necessary. We *can* change the course of the ship from one that leads to separateness to one that sails toward the oneness of all life.

ABOUT THE CONTRIBUTORS

Irene van Lippe-Biesterfeld is a writer, mother of four, and teacher of intuitive development who has long worked in the field of conservation and nature preservation. Her most recent project involves the restoration and preservation of a large tract of land in South Africa with an eye to it becoming a full-fledged nature preserve. She is founder and president of the Lippe-Biesterfeld Nature College Foundation and has developed its course, "Dialogue with Nature." The author will donate her proceeds from the sale of *Science, Soul, and the Spirit of Nature* to the Nature College Foundation to help further its mission of promoting an understanding of the spirituality in nature and our role as both guardians and co-participants in nature's diverse and interdependent web. Please visit the Lippe-Biesterfeld Nature College Foundation at www.naturecollege.org.

Jessica van Tijn holds degrees in international relations and communications from the University of Amsterdam and a postgraduate degree in international relations from the Chingendael Institute in The Hague. After working in both Dutch public radio and television, she moved to Mexico in 1998 to found Special Eyes Productions (www.specialeyes.nl), which produces radio programs and documentaries. Currently, she writes for several principally Dutch magazines and newspapers and is a correspondent for the daily Dutch radio news program *Wereldnet*. She lives in Mexico, where her daughter is enjoying a trilingual (Dutch, Spanish, and English) upbringing.

Hans Andeweg began as an agricultural biologist before joining the Institute for Resonance Therapy (IRT) in Germany in 1989. Upon leaving the IRT, he wrote *In Resonance with Nature,* on which his ecological energy-balancing method is based, and cofounded the Center for ECOtherapy with his wife.

Masaru Emoto was born in Japan and is a graduate of the Yokohama Municipal University and of the Open International University as a doctor of alternative medicine. He is well known for his photographic research into the study of the effects of thoughts and feelings on water and is the author of *Messages from Water 1 and 2* and *The Hidden Messages in Water.*

Jane Goodall (right), famous for her studies of chimpanzees, has established the Gombe Stream Research Centre in Tanzania; founded the Jane Goodall Institute for Wildlife Research, Education, and Conservation; and created the Roots & Shoots program to inspire young people to implement local projects that promote care for animals, the environment, and the human community.

Denise Linn (left) is an international lecturer, healer, feng shui practitioner, and author of *Sacred Space, Soul Coaching,* and *Feng Shui for the Soul.* She conducts soul coaching seminars at Summerhill Ranch in California.

Patricia Mische (left) is cofounder of Global Education Associates, Lloyd Professor of Peace Studies and world law at Antioch College, and author of more than one hundred articles and several books, including *Star Wars and the State of Our Souls* and *Toward a Global Civilization? The Contribution of Religions*. She regularly gives talks on topics related to peace, human rights, women, ecological security, and world order, and is the recipient of numerous awards for her contributions to the field of global and peace education.

Credo Vusamazulu Mutwa is an oral historian and traditional Zulu healer *(sanusi)* who resides in Pretoria, South Africa, near Johannesburg, where he continues to heal people, sculpt, paint, and teach Zulu lore. He is the author of several books, including *Zulu Shaman* and *Indaba, My Children.*

Arne Naess, the founder of deep ecology and one of Norway's best-known philosophers, is professor emeritus at the University of Oslo and has been working with the Center for Development and the Environment (SUM: Senter for utvikling og miljø) in Norway since 1991.

Gareth Patterson, a protégé of the Adamsons, whose story was told in *Born Free,* has pioneered the practice of returning lions to the wild. He is an environmentalist, a speaker, and the author of seven books on lions and his experiences in southern Africa and Kenya.

Matthijs G. C. Schouten is associate professor of restoration ecology at the University of Wageningen, the Netherlands, and associate professor of nature and landscape conservation at the Universities of Cork and Galway in Ireland. He is the senior ecologist with the National Forest Service of the Netherlands and is also a visiting professor at both University College Cork in Ireland and the National University of Ireland, Galway.

Rupert Sheldrake is a biologist, a former research fellow of the Royal Society at Cambridge, a current fellow of the Institute of Noetic Sciences near San Francisco, and an academic director and visiting professor at the Graduate Institute in Connecticut. He received his Ph.D. in biochemistry from Cambridge University and was a fellow of Clare College, Cambridge University, where he carried out research on the development of plants and the ageing of cells. He is the author of more than seventy-five scientific papers and ten books, including *Dogs That Know When Their Owners Are Coming Home; A New Science of Life; The Presence of the Past; Chaos, Creativity, and Cosmic Consciousness; The Rebirth of Nature;* and *Seven Experiences That Could Change the World.*

Rigoberta Menchú Tum, a 1992 Nobel Peace Prize recipient, was raised in the Quiché branch of the Mayan culture and is a leading advocate for Indian rights and ethno-cultural reconciliation.

James Wolfensohn, the ninth president of the World Bank (succeeded by Paul Wolfowitz on June 1, 2005), has made sustainable poverty reduction the World Bank's overarching mission for the past ten years. Prior to joining the World Bank, he established his career as an international investment banker with a parallel involvement in development issues and the global environment. He is currently working as a United Nations special envoy to the Israeli pullout from the Gaza Strip.

BOOKS OF RELATED INTEREST

Science and the Akashic Field
An Integral Theory of Everything
by Ervin Laszlo

The Secret Teachings of Plants
The Intelligence of the Heart in the Direct Perception of Nature
by Stephen Harrod Buhner

The Rebirth of Nature
The Greening of Science and God
by Rupert Sheldrake

The Presence of the Past
Morphic Resonance and the Habits of Nature
by Rupert Sheldrake

The Universe Is a Green Dragon
A Cosmic Creation Story
by Brian Swimme, Ph.D

Green Psychology
Transforming Our Relationship to the Earth
by Ralph Metzner, Ph.D.

The Nature of Things
The Secret Life of Inanimate Objects
by Lyall Watson

Radical Knowing
Understanding Consciousness through Relationship
by Christian de Quincey

Inner Traditions • Bear & Company
P.O. Box 388
Rochester, VT 05767
1-800-246-8648
www.InnerTraditions.com

Or contact your local bookseller